Make: Action

WITHDRAWN

*Movement, Light, and Sound with
Arduino and Raspberry Pi*

Simon Monk

MAKER MEDIA™
SAN FRANCISCO, CA

Make: Action

by Simon Monk

Published by Maker Media, Inc., 1160 Battery Street East, Suite 125, San Francisco, CA 94111.

Maker Media books may be purchased for educational, business, or sales promotional use. Online editions are also available for most titles (*http://safaribooksonline.com*). For more information, contact O'Reilly Media's institutional sales department: 800-998-9938 or *corporate@oreilly.com*.

Editor: Roger Stewart

Production Editor: Nicole Shelby

Copyeditor: Jasmine Kwityn

Proofreader: Kim Cofer

Indexer: Ellen Troutman

Interior Designer: David Futato

Cover Designer: Karen Montgomery

Illustrator: Rebecca Demarest

January 2016:   First Edition

Revision History for the First Edition

2016-02-02:   First Release

See *http://oreilly.com/catalog/errata.csp?isbn=9781457187797* for release details.

978-1-457-18779-7

[TC]

Table of Contents

Introduction 1

The Arduino and Raspberry Pi make it easier than ever for a hobbyist to get into the world of electronics. Perhaps you want to set up a DIY home automation system so that you can control your lights and heating over your WiFi network, or simply control some motors.

This book will show you how to use the popular Raspberry Pi and Arduino platforms so that your Pi or Arduino can make and control movement, light, and sound.

Arduino and Pi

Although the Arduino and Raspberry Pi are both small, credit card–sized circuit boards, they are actually quite different devices. The Arduino is a very simple microcontroller board that does not run an operating system of any sort, whereas the Raspberry Pi is a tiny computer that runs Linux and also happens to be able to interface to external electronics.

Raspberry Pi

If you are new to electronics, but comfortable using computers, then the Raspberry Pi is going to be the more familiar device. The Raspberry Pi (Figure 1-1) really is a very small version of a regular computer running Linux. It has USB ports for you to attach a keyboard and mouse as well as HDMI video output to connect to a monitor or TV and audio output.

Figure 1-1 *A Raspberry Pi 2*

The Raspberry Pi has an Ethernet port to connect it to your network and will also accept USB WiFi adapters. Power is supplied to the Raspberry Pi using a microUSB socket.

A microSD card is used for storage rather than a conventional disk drive. This card contains both the operating system and all your documents and programs.

The Raspberry Pi was created in the UK, primarily to serve as a low-cost computer to help with teaching computer basics, particularly Python programming, to school kids. In fact, the name Pi is said to be derived from the *Py* of *Python*.

A few things set the Raspberry Pi apart from a regular desktop or laptop computer running Linux:

- It costs just $40 (a cut-down Raspberry Pi called the model A+ is also available for a lower price and the model zero for an even lower price).

- It uses under 5 watts of power.

- The Raspberry Pi has a double row of general-purpose input/output (GPIO) pins that allow you to connect electronics directly to it (the pins can be seen in the upper-left part of Figure 1-1. From these pins, you can control LEDs, displays, motors, and all the different types of output devices that you will work with later in this book.

In addition, the Raspberry Pi can also be connected to the Internet using WiFi or a LAN cable, making it suitable for Internet of Things projects (Chapter 16).

The specs for the Raspberry Pi 2 (the latest and best version at the time of writing) are as follows:

- ARM v7 900 MHz quad-core processor
- 1 GB DDR2 memory
- 10.100 BaseT Ethernet
- 4 USB 2.0 ports
- HDMI video out
- Camera interface socket
- 40-pin GPIO header (all pins operate at 3.3V)

If you are new to Raspberry Pi, there is a primer to get you up and running with the hardware and also the Python programming language in Chapter 3.

Arduino

There is quite a wide range of different Arduino models available. This book concentrates on the most widely used and popular Arduino model, the Arduino Uno (Figure 1-2). The Arduino is a little cheaper than a Raspberry Pi—you can purchase an Arduino Uno for around $25.

Figure 1-2 *An Arduino Uno Revision 3*

If you are accustomed to working on a regular computer, the Arduino's specifications will probably seem grossly inadequate. It has just 34 KB of memory (of various types). That means that the Raspberry Pi has roughly 30,000 times as much memory, without even including the flash memory of the Pi's SD card! What's more, the Arduino Uno has a processor clocked at just 16MHz. You cannot attach a keyboard, mouse, or monitor to the Arduino, and it does not run an operating system.

You may wonder how this little device can actually do anything useful. The secret to the Arduino's usefulness lies in its very simplicity. There is no operating system to boot up, or other interfaces that you may not need in a project that would simply add cost and consume power.

Whereas the Raspberry Pi is a general computer, the Arduino concentrates on doing one thing well—connecting to and controlling electronics.

To program an Arduino, you need a regular computer (you can even use a Raspberry Pi if you want). On the computer you choose, you'll need to run an integrated development environment (IDE), which allows you to write a program to be downloaded to the built-in flash memory on the Arduino.

The Arduino can only run one program at a time, and once it is programmed, it will remember that program and automatically run it as soon as it is powered up.

Arduinos are designed to accept *shields*, which are boards that plug on top of the Arduino's input/output sockets and add extra hardware features, such as various types of display, as well as Ethernet and WiFi adapters.

You program an Arduino using the C programming language (you can find out more about programming and using an Arduino in Chapter 2).

Choosing a Device: Arduino or Pi?

One of the reasons this book explains how to connect electronics to both Arduino and Raspberry Pi is that some projects are better suited to a Raspberry Pi and some to an Arduino. Other boards that lie in between these extremes are generally close enough to either an Arduino or Raspberry Pi that this book will be of use in working out how to use them.

When embarking on a new project, my rule of thumb is to use an Arduino by default. However, if the project has one of the following requirements, then a Raspberry Pi is probably a better choice:

- Internet or network connectivity
- The need for a large screen
- The need to attach keyboard and mouse
- The need for USB peripherals such as a web cam

With some effort and expense it's possible to expand the Arduino with shields to cover most of the preceding requirements. However, it will be harder to get everything working if you go this route, as none of these are native features of the Arduino in the same way as they are for the Pi.

Good reasons for using an Arduino over a Raspberry Pi include the following:

Cost
> An Arduino Uno is cheaper than a Raspberry Pi 2.

Startup time
> An Arduino does not need to wait while an operating system boots up. There is a small delay of about a second while it checks to see if a new program is being uploaded and then it's up and running.

Reliability
> An Arduino is intrinsically a much simpler and tougher device than a Raspberry Pi and doesn't have the overhead of an operating system.

Power consumption
> An Arduino uses about 1/10th of the power of a Raspberry Pi. If you need a battery- or solar-powered solution, then the Arduino is a better choice.

GPIO output current
> A Raspberry Pi's GPIO pin should only be used to supply a maximum of around 16 mA. On the other hand, an Arduino pin is rated at 40 mA. So, in some cases, you can connect something (say a bright LED) to an Arduino directly, in a way that you couldn't with a Raspberry Pi.

The Arduino and the Raspberry Pi both are great devices to base a project on, and to some extent, the choice of which device to use will also be a matter of personal preference.

One important thing to remember when attaching external electronics to a Raspberry Pi is that it operates at 3.3V, rather than the 5V of Arduino. Connecting 5V to one of the Raspberry Pi's GPIO pins is likely to damage or destroy the GPIO pin or even the whole Raspberry Pi.

Alternatives

The Arduino Uno and Raspberry Pi sit at each end of the range of devices that can be used to control things. As you might expect, the market has produced a whole host of other devices that sit in between these two extremes, in some cases trying to give the best of both worlds.

New devices are popping up all the time. The open source nature of Arduino has led to lots of variations on it, with designs for specific niches, such as controlling drones or interfacing to wireless sensors.

Figure 1-3 shows a spread of some of the most popular devices in this area.

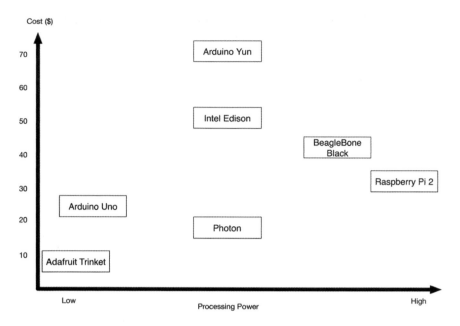

Figure 1-3 *Embedded platforms*

Below the Arduino Uno, both in price and performance, is the Adafruit Trinket. This interesting board only has a few GPIO pins, but is otherwise fairly Arduino compatible. It's worth considering for a project that may just have one or two inputs or outputs.

There is a middle ground of products comprising the Arduino Yun, Intel Edison, and Photon that all have built-in WiFi capabilities and are intended for use with Internet of Things (IoT) projects (see Chapter 16). Of these, the Photon probably represents the best value. All three of these devices are programmed using Arduino C, so what you learn about using the Arduino will also apply to these boards.

The BeagleBone Black is very similar in concept to the Raspberry Pi. It too is a single-board computer and although the current BeagleBone Black version has fallen behind the Raspberry Pi 2 in terms of raw power, the BeagleBone Black does have some advantages over the Raspberry Pi. It has more GPIO pins and also has some pins that can act as analog inputs, a feature lacking in the Raspberry Pi 2. The BeagleBone Black can either be programmed in Python, in a similar way to the Raspberry Pi, or in JavaScript.

Summary

This chapter provided a brief introduction to the Arduino and the Raspberry Pi. We discussed the advantages and disadvantages that each of these boards offer, and also looked at some alternatives. The next two chapters will get you started with using and programming the Arduino and then the Raspberry Pi.

If you have used Arduino and Raspberry Pi before, you might like to jump ahead to Chapter 4 and use the Arduino and Raspberry Pi to Make some Action! You can always return to Chapter 2 and Chapter 3 if you need to.

Arduino | 2

This chapter is a modified version of the Arduino primer initially published as an appendix of my book *The Maker's Guide to the Zombie Apocalypse* from NoStarch Press and is used here with their kind permission.

If you are new to Arduino, this chapter will get you started with this great little board.

What Is an Arduino?

There are various types of Arduino boards, but by far the most common and the one that is used for all the projects in this book is the Arduino Uno (see Figure 2-1).

The Arduino Uno has gone through a number of revisions. The Arduino Uno shown in Figure 2-1 is a revision 3 (R3) board (the latest at the time of writing).

Let's start our tour of the Arduino with the USB socket, which serves several purposes: it can be used to provide power to the Arduino, for programming the Arduino from your computer, and finally, as a communications link.

The little red button next to the USB socket is the Reset button. When you press this, the Arduino will restart and run the program that is installed on it.

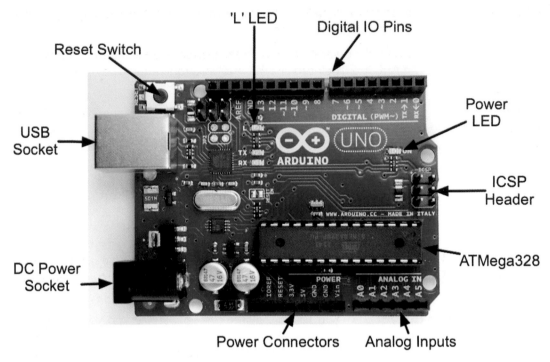

Figure 2-1 *An Arduino Uno R3*

Along both the top and bottom edges of the Arduino, there are connection sockets to which electronics can be attached. The digital inputs and outputs, identified by a number between 0 and 13, can be seen at the top of Figure 2-1. Each pin can be configured in your programs to be either an input or an output. You might connect a switch to a digital input, and the digital input will be able to tell if the switch is pressed or not. Alternatively, you might connect an LED to a digital output and by changing the digital output from low to high, you can turn the LED on. In fact there is one LED built onto the board called the "L" LED that is connected to digital pin 13.

Beneath the digital I/O pins, there is a power LED that simply indicates that the board is powered. The ICSP (In-Circuit Serial Programming) header is only for advanced programming of the Arduino without using the USB connection. Most users of Arduino will never use the ICSP header.

It is worth highlighting the ATMega328 (the brains of the Arduino). The ATMega328 is a microcontroller IC (Integrated Circuit). This chip stores the program that the Arduino will run (it contains 32 KB of flash memory).

Below the ATMega328, there is a row of analog input pins labeled A0 to A5. Whereas digital inputs can only tell if something is on or off, analog inputs can measure the voltage at the pin (as long as the voltage is between 0 and 5V). Such a voltage could be from a sensor of some sort. If you run short of digital inputs and outputs, then these analog input pins can also be used as digital inputs and outputs.

Next to this, there is a row of power connectors that can be used as alternative ways of supplying power to the Arduino. They can also be used to supply power to other electronics that you build.

The Arduino also has a DC power jack. This can accept anything between 7V and 12V DC and uses a voltage regulator to provide the 5V that the Arduino operates on. The Arduino will automatically accept power from the USB socket and power from the DC connector depending on which is connected.

Installing the Arduino IDE

The Arduino is not quite what you would expect from a computer. It has no operating system, and you do not connect a keyboard, monitor, or mouse to it. It only ever runs a single program and you have to upload that program to its flash memory using a proper computer. You can reprogram the Arduino as many times as you like (well, many thousands of times).

In order to program the Arduino, you need to install the Arduino IDE on your computer. The cross-platform nature of this software—the Arduino IDE can be run on Windows, Mac, and Linux—is one of the reasons for the Arduino's great popularity. In addition, the Arduino IDE allows you to program the Arduino over a USB connection without any need for special programming hardware.

To install the Arduino IDE for your platform, download the software and then follow the instructions on the Arduino website (*http://arduino.cc/en/Guide/HomePage*).

Note that Windows and Mac users will need to install USB drivers so that the Arduino IDE can communicate with the Arduino itself.

Once everything is installed, run the Arduino IDE. Figure 2-2 shows the Arduino IDE window.

The Upload button will, as the name suggests, upload the sketch to the Arduino board. Before it uploads the sketch, it converts the textual programming code into executable code for the Arduino. If there are errors, they will be displayed in the Log area. The Verify button does the same thing but without taking the final step of uploading the program to the board.

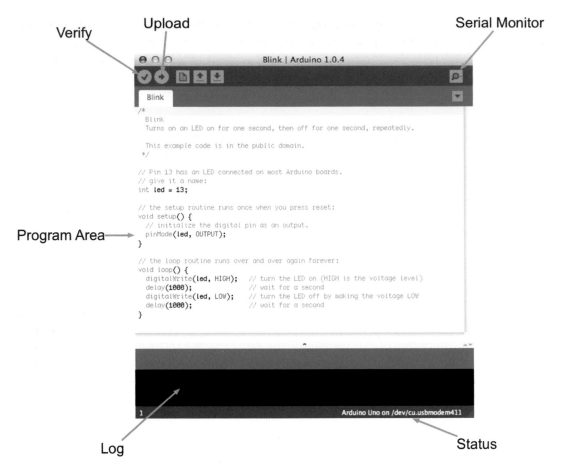

Figure 2-2 *The Arduino IDE*

The Serial Monitor button opens up the Serial Monitor window, which is used for communicating with the Arduino. You will use the Serial Monitor in many of the experiments in this book, as it is a great way of sending commands to an Arduino from your computer. The Serial Monitor allows two-way communication, which means that you can send text messages to the Arduino and also receive responses from it.

The Status area at the bottom of the window will tell you the type of Arduino and its corresponding serial port through which it will be programmed when the Upload button is pressed. The port shown in Figure 2-2 (*/dev/cu.usbmodem411*) is of the type you would expect to see when using a Mac or Linux computer. If you are using a Windows computer to program the Arduino, then this will be COM4, or COM followed by a number that Windows allocated to the Arduino when you connected it.

Last but not least, the main part of the Arduino IDE is the Program Area where you type the program code that you want to upload to the Arduino.

In the world of Arduino, programs are called *sketches*, and the File menu of the Arduino IDE allows you to Open and Save sketches as you would a document in a word processor. The File menu also has an Examples sub-menu from which you can load Arduino's built-in example sketches.

Uploading a Sketch

To test out your Arduino board, and to make sure the Arduino IDE is properly installed, start by opening the example sketch called Blink. You will find this by navigating to File –› Examples –› 01. Basics. (The Blink sketch is the example that is loaded in Figure 2-2).

Use a USB lead to attach your Arduino to the computer you are going to use to program the Arduino. You should see the power LED of the Arduino light up and a few of the other LEDs flicker as it is plugged in.

Now that the Arduino is connected, you need to tell the Arduino both the type of board being programmed (Arduino Uno) and the serial port that it is connected to. Set the board type by navigating to Tools –› Board –› Arduino Uno.

Set the serial port by navigating to Tools –› Serial Port. If you are using a Windows computer there are not likely to be many options there. In fact, you may just find the option COM4. On a Mac or Linux computer, there are generally a variety of USB devices listed and it can be difficult to determine which one is your Arduino board. You should see an item in the list that starts *dev/tty.usbmodemNNN* where *NNN* is a number. In Figure 2-3, the Arduino attached to my Mac has been selected.

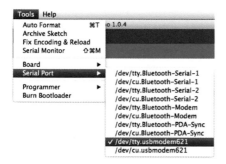

Figure 2-3 *Selecting the Arduino serial port*

If your Arduino does not show up in the list, then this usually points to a problem with the USB drivers. If so, try reinstalling them.

Once you are ready to upload the sketch to the Arduino, press the Upload button. Messages should start to appear in the Log area of the window and then after a few seconds, the LEDs labeled "TX" and "RX" on the Arduino should start flickering as the program is uploaded onto the board.

If everything goes according to plan, when the upload is complete, you should see a message something like the one shown in Figure 2-4.

Figure 2-4 *A successful upload*

This message tells you that the sketch has uploaded, and that it has used 1,084 bytes of the 32,256 bytes available.

After the sketch has finished uploading, you will notice that the built-in "L" LED on the Arduino will be slowly blinking on and off. This is your newly uploaded Blink sketch living up to its name.

The Book Code

All of the software for this book—both Arduino sketches (as Arduino programs are called) and Python programs for the Raspberry Pi—is available from the book's GitHub page (*https://github.com/simonmonk/make_action*).

To get the files onto your Mac, Linux, or Windows computer, click the Download ZIP button, which is found in the bottom right of the GitHub page.

This will download a ZIP file that you can save on your desktop or other convenient location. When you expand the ZIP archive, it will unpack a directory called *make_action-master/*. You will find the Arduino code in a directory called *arduino/*. Inside the *arduino/* directory are two subdirectories, *experiments/* and *projects/*.

Each experiment or project program is held in a directory of its own and within that there is usually a single file that is the actual program. For example, within the directory *experiments/*, you will find the directory *ex_01_basic_motor_control/*, which contains the single file *basic_motor_control.ino*. If you already have the Arduino IDE installed, then opening this file will open it in the Arduino IDE.

An alternative way of gaining access to these sketches is to copy both the *experiments* and *projects* folders into your Arduino sketchbook directory, which is called *Arduino/* and will be in your normal documents folder (such as *My Documents/* on Windows or *Documents/* on a Mac).

If the files are copied to the *sketchbook/* directory, then you will be able to open them by navigating to File –› Sketchbook in the Arduino IDE.

Programming Guide

This section contains an overview of the main commands to help you understand the sketches used in the book. If you are new to programming and want to learn Arduino C, you should consider reading my book *Programming Arduino: Getting Started with Sketches*.

Setup and Loop

All Arduino sketches must include a setup() function and a loop() function (functions are blocks of program code that do something). To see how the setup() and loop() work, let's dissect the Blink example that we uploaded to the Arduino:

```
int led = 13;
// the setup routine runs once when you press reset:
void setup() {
  // initialize the digital pin as an output.
  pinMode(led, OUTPUT);
}

// the loop routine runs over and over again forever:
void loop() {
  digitalWrite(led, HIGH);    // turn the LED on (HIGH is the voltage level)
  delay(1000);                // wait for 1 second
  digitalWrite(led, LOW);     // turn the LED off by making the voltage LOW
  delay(1000);                // wait for 1 second
}
```

You will notice that quite a lot of the text of the sketch has a // in front of it. This indicates that the rest of the text after // until the end of the line is to be treated as a comment. It's not program code, it's just documentation that tells the person looking at the program what is going on.

As the comments say, the lines of code inside the setup() function are run just once (or more precisely every time power is applied to the Arduino or the Reset button is pressed). So, you use setup() to do all the things you need to do just once when the program starts. In the case of Blink, this just means specifying that the LED pin is set to be an output.

The commands inside the loop() function will be run over and over again—that is, as soon as the last of the command lines in loop() has done its business, it will start over on the first line again.

I have skipped over what the commands inside setup() and loop() actually do in Blink, but don't worry, I will discuss them soon.

Variables

Variables are a way of giving names to values. The first line of Blink (ignoring comments) is as follows:

```
int led = 13;
```

This defines a variable called led and gives it an initial value of 13. The value 13 is chosen because this is the name of the Arduino pin that the L LED is connected to and int is the type of variable. The word int is short for *integer* and means a whole number (no decimal points).

Although it's not required to use a variable name for every pin that you use, it is a good idea to do so, because it makes it easier to see what the pin is used for. Also, should you want to use a different pin, you only need to change the value of the variable in one place where you define it.

You may have noticed that in the sketches that accompany this book, when declaring variables like this that define a pin to be used, the line has const at the front of it like this:

```
const int led = 13;
```

The const keyword tells the Arduino IDE that the variable is not really a variable, but is a constant; in other words, its value is never going to change from 13. This does result in slightly smaller and faster running sketches and is generally considered to be a good habit to get into.

Digital Outputs

The Blink sketch is a good example of a digital output. Pin 13 is configured to be an output in the setup() function by this line (the variable led having earlier been defined to be 13):

```
pinMode(led, OUTPUT);
```

This is in the setup() function because it only needs to be done once. Once the pin is set to be an output, it will stay as an output until we tell it to be something else.

Turning the LED on and off to make it blink needs to happen more than once, so the code for this goes inside:

```
digitalWrite(led, HIGH);    // turn the LED on (HIGH is the voltage level)
delay(1000);                // wait for 1 second
digitalWrite(led, LOW);     // turn the LED off by making the voltage LOW
delay(1000);                // wait for 1 second
```

The digitalWrite() has two parameters (inside parentheses and separated by a comma). The first parameter is the Arduino pin to write to and the second parameter is the value to be written to the pin. So, a value of HIGH will set the output to 5V (which turns on the LED) and a value of LOW will set the pin to 0V turning the LED off.

The delay() function pauses the program for the amount of time (in milliseconds) specified as a parameter. There are 1,000 milliseconds in 1 second, so each of the delay() functions shown here will pause the program for 1 second.

In "Experiment: Controlling an LED" on page 46 you will use a digital output connected to an external LED and make that blink rather than the built-in LED.

Digital Inputs

Being mostly concerned with outputs rather than inputs, this book primarily uses digitalWrite(). However, you should also know about digital inputs, which can be used to attach switches and sensors to an Arduino.

You can set an Arduino pin to be a digital input using the pinMode() function. The following example would set pin 7 to be an input. You could of course use a variable name in place of 7.

Just as you would with an output, you define an input pin in the setup() function, as it is rare to change the mode of a pin once the sketch is running:

```
pinMode(7, INPUT)
```

Having set the pin to be an input, you can then read the pin to determine if the voltage at that pin is closer to 5V (HIGH) or closer to 0V (LOW). In this example, the LED will be turned on if the input is LOW at the time it is tested (once lit, the LED will stay lit, as there is nothing in the code to turn it off again):

```
void loop()
{
  if (digitalRead(7) == HIGH)
  {
    digitalWrite(led, LOW)
  }
}
```

Suddenly, the code has started to get a little complex, so we'll go over it one line at a time.

On the second line, we have a { symbol. Sometimes this is put on the same line as loop() and sometimes on the next line. This is just a matter of personal preference; it has no effect on the running of the code. The { symbol marks the start of a block of code that ends with its matching } symbol. It is a way of grouping together all the lines of code that belong to the loop function.

The first of these lines uses the if statement. Immediately after the word "if" is a condition. In this case the condition is (digitalRead(7) == HIGH). The double equals sign (==) is a way of comparing the two values on either side of it. So, in this case, if pin 7 is HIGH, then the block of code surrounded by { and } after if will run, otherwise it won't. Lining up the { and } makes it easier to see which } belongs to which {.

We have already seen the code to be run if the condition is true. This is the digitalWrite() function that we used to turn the LED on.

In this example, it is assumed that the digital input is firmly HIGH or LOW. If you are using a switch connected to a digital input, then all that switch can do is close a connection. Typically, this will mean connecting the digital input to GND (0V). If the switch's connection is open, then the digital input will be said to be floating. In other words, electrically it's not connected firmly to anything. The input will pick up electrical noise and can often alternate between high and low. To prevent this undesirable behavior, a pull-up resistor is normally used, as shown in Figure 2-5.

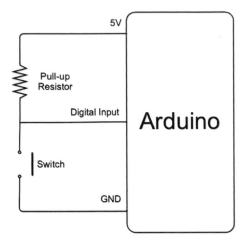

Figure 2-5 *Using a pull-up resistor with a digital input*

When the switch is not closed (as shown in Figure 2-5) the resistor pulls the input pin up to 5V. When the switch is closed by pressing the button, the weak pulling up of the input is overwhelmed by the switch connecting the digital input to GND.

Although you can use an actual pull-up resistor, Arduino inputs have built-in pull-up resistors of about 40kΩ in value that are enabled if you set the pin mode to be INPUT_PULLUP rather than just INPUT. The following code demonstrates how you would set the pin mode of a digital input to be used with a switch, without having to use an external pull-up resistor shown in Figure 2-5:

```
pinMode(switchPin, INPUT_PULLUP);
```

Analog Inputs

Analog inputs allow you to measure a voltage between 0 and 5V on any of the special analog input pins of the Arduino labeled A0 to A5. Unlike digital inputs and outputs, you do not need to use the pinMode() in setup when using an analog input.

To read the value of an analog input, use the analogRead() function, supplying the name of the pin you want to read as a parameter. Unlike digitalRead(), analogRead() returns a number rather than just true or false. The number that you get back from analogRead() is a number between 0 and 1,023. A result of 0 means 0V and 1,023 means 5V. To convert the number you get back from analogRead() to an actual voltage, you need to multiply the number by 5 and then divide it by 1,023—or you can simply divide it by 204.6. Here's how you would do it in Arduino C:

```
int raw = analogRead(A0);
float volts = raw / 204.6;
```

The variable `raw` is an `int` (whole number) because the reading from an analog input is always a whole number. However, to scale the raw reading to be a decimal number, the variable needs to be of type `float` (floating point).

You can connect various sensors to analog inputs—for example, later in the book, you will learn how to use an analog input with a photoresistor light sensor (see "Project: Arduino House Plant Waterer" on page 114) and how to connect a variable resistor (see "Project: A Thermostatic Beverage Cooler" on page 246).

Analog Outputs

Digital outputs allow you to turn something (for example, an LED) on and off. But if you want to control the power to something in a graduated way, you need to use an analog output. An analog output is useful for controlling the brightness of an LED or the speed of a motor, for instance. As you might expect, this book will use this feature quite a lot.

Not all the pins of an Arduino Uno are capable of acting as an analog output—you must use D3, D5, D6, D9, D10, or D11. These pins are marked with a little ~ next to the pin on the Arduino itself.

To control an analog output, use the `analogWrite()` function with a number between 0 and 255 as the parameter. A value of 0 means 0V (completely off), and a value of 255 is fully on.

It is tempting to think of an analog output as being a voltage between 0 and 5V, and if you attach a volt meter between a pin being used as an analog output and GND, the voltage will indeed seem to change between 0 and 5V as you change the value that you use in your `analogWrite()`. In fact, things are a little more complex than that. Figure 2-6 shows what is really going on with this kind of output, which is called pulse-width modulation (PWM).

An analog output pin generates 490 pulses per second on all the analog output–capable pins (except D5 and D6, which operate at 980 pulses per second). The width of the pulses are varied. The larger the proportion of the time that the pulse stays high, the greater the power delivered to the output, and hence the brighter the LED or faster the motor.

The reason that a volt meter reports this as a change in voltage is that the volt meter cannot respond fast enough, and therefore does a kind of averaging to produce a voltage that appears to vary smoothly.

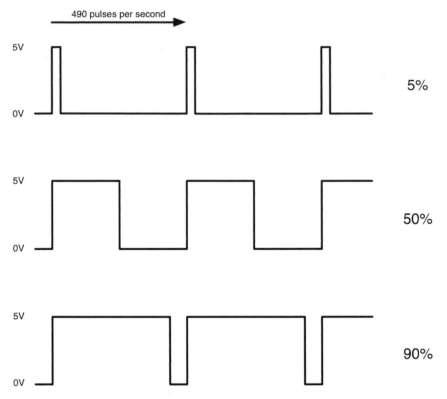

Figure 2-6 *Analog outputs pulse-width modulation*

If/Else

As you'll recall from "Digital Inputs" on page 17, we used an `if` statement to perform a specified action if a certain condition is true. To gain even greater control over the flow of code, you can use `if/else`, which will perform one set of code if the condition is true and a different set of code if it is false.

The following example would turn the pin LED on and off depending on whether an analog reading is greater than 500 or less than or equal to 500:

```
if (analogRead(A0) > 500)
{
  digitalWrite(led, HIGH);
}
else
{
  digitalWrite(led, LOW);
}
```

So far we have met two types of comparison operators: == (equal to) and > (greater than). There are actually some more comparisons you can make:

- <= (less than or equal to)

- >= (greater than or equal to)

- != (not equal to)

In addition to comparing only two values, you can also make more complicated comparisons using && (and) and ||(or). For example, if you wanted to only turn the LED on if the reading was between 300 and 400, you would write the following:

```
int reading = analogRead(A0);
if ((reading >= 300) && (reading <=400))
{
  digitalWrite(led, HIGH);
}
else
{
  digitalWrite(led, LOW);
}
```

Loops

A loop allows code to be repeated a specified number of times or until some condition changes. To accomplish this, you either use a for loop or a while loop: for loops are better for doing something a fixed number of times, and while loops are better for doing something while some condition is true.

The for loop in the following example will make the LED blink 10 times (note that the for loop is in setup() rather than loop(), because if you put it in loop() it would blink another 10 times after it had finished the first 10 and so on, which is not the desired effect):

```
for (int i = 0; i < 10; i++)
{
  digitalWrite(led, HIGH);
  delay(1000);
  digitalWrite(led, LOW);
  delay(1000);
}
```

If you wanted to keep an LED blinking for as long as a button connected to a digital input is pressed, then you would use a while loop:

```
while (digitalRead(9))
{
  digitalWrite(led, HIGH);
  delay(1000);
  digitalWrite(led, LOW);
  delay(1000);
}
```

This code assumes that pin D9 is connected to a switch (as shown in Figure 2-5).

Functions

Functions can cause a lot of confusion to those that are new to programming. Perhaps the easiest way to think of functions is as ways of grouping together some lines of code and giving them a name so that it is easy to use them over and over again.

If you were to look at some of the internal workings of Arduino, you would find that the built-in functions such as digitalWrite() are actually more than a little complicated. For example, here is the code for digitalWrite() (don't worry what it does, just be glad that you don't have to type all that code in every time you want to change pins from high to low):

```
void digitalWrite(uint8_t pin, uint8_t val)
{
    uint8_t timer = digitalPinToTimer(pin);
    uint8_t bit = digitalPinToBitMask(pin);
    uint8_t port = digitalPinToPort(pin);
    volatile uint8_t *out;

    if (port == NOT_A_PIN) return;

    // If the pin that support PWM output, we need to turn it off
    // before doing a digital write.
    if (timer != NOT_ON_TIMER) turnOffPWM(timer);

    out = portOutputRegister(port);

    uint8_t oldSREG = SREG;
    cli();

    if (val == LOW) {
        *out &= ~bit;
    } else {
        *out |= bit;
    }

    SREG = oldSREG;
}
```

By giving that big chunk of code a name, we can just refer to it by name to make use of it.

In addition to built-in functions, like digitalWrite, you can create your own functions to lump together things you use. For example, you could create a function that will blink the LED a number of times specified as a parameter. The pin to blink could also be specified as a parameter. The following sketch illustrates this, by making a function called blink and calling it during startup() so that the Arduino "L" LED blinks 5 times after a reset:

```
const int ledPin = 13;
void setup()
{
  pinMode(ledPin, OUTPUT);
  blink(ledPin, 5);
}
```

```
void loop() {}

void blink(int pin, int n)
{
  for (int i = 0; i < n; i++)
  {
    digitalWrite(ledPin, HIGH);
    delay(500);
    digitalWrite(ledPin, LOW);
    delay(500);
  }
}
```

The setup() function sets the ledPin to be an output and then calls the function blink, passing it the pin that should be blinked followed by the number of times to blink it. The loop() function is empty and does nothing, but the Arduino IDE will still insist that it is there.

The blink function itself begins with the word void, which indicates that the function does not return anything—in other words, you cannot assign the result of calling that function to a variable, as you might want to do if the function performed some kind of calculation. Following that, we see the name of the function (blink) and then the parameters that the function takes enclosed within parentheses and separated by commas. When you define a function, you must specify the type of each of the parameters (that is, whether they are an int, float, or something else). In this case, both the pin to blink (pin) and the number of times to blink (n) are ints (whole numbers).

As with most programming languages, the C programming language has the concept of global and local variables. *Global variables* (such as ledPin in the preceding example) can be used from anywhere in the program. On the other hand, *local variables* such as parameters to functions (pin and n in this example) and even i inside the for loop are only accessible within the function in which they are defined.

So, in setup(), the line blink(ledPin, 5) passes the global variable ledPin into the function blink, where it will be assigned to the local variable pin. You might wonder why you should do this. The answer is that by passing in the pin to blink, we make the blink function general purpose, so it can be used to flash any pin we tell it to, rather than just ledPin.

In the body of the blink function, we have a for loop that will repeat the digitalWrite() and delay() functions inside it n times. So, if n is 3, the LED will blink 3 times.

Summary

In this chapter, you have learned how to install the Arduino IDE and get started with a few basic Arduino programming commands, and looked at how your code can control the Arduino's input/output pins.

If you are new to programming, some of the concepts described in this chapter will take some time to sink in. One of the best ways to learn how to program is to look at existing examples and then alter them to get a feeling for what everything does.

All the program code used in this book is available for download via the book's GitHub repository (*https://github.com/simonmonk/make_action*), so you do not have to write any of your own programs to make use of this book, although you will probably find yourself wanting to use the experiments and projects as a basis for your own project ideas.

In the next chapter, you will find a similar primer to this one, but for the Raspberry Pi.

Raspberry Pi | 3

This chapter will help you get started with your Raspberry Pi. If you are already familiar with the Raspberry Pi and have tried a little Python programming on the Raspberry Pi, then you might like to jump ahead to Chapter 4.

What Is a Raspberry Pi?

A Raspberry Pi is a single-board computer, complete with Linux operating system, USB connections for keyboard and mouse, and an HDMI connector for attaching a monitor.

There are actually a few variations on the Raspberry Pi, some historic and no longer manufactured. All of the Raspberry Pi models are broadly compatible, and you should not encounter any problems with the examples in this book, even if you have an old Raspberry Pi.

Figure 3-1 shows a Raspberry Pi 2 model B. On the righthand side of the board you will find four USB ports. These are useful for attaching a keyboard and mouse, as well as other peripherals, such as printers, scanners, and Flash drives.

Below the USB ports, you will find an RJ45 Ethernet socket that allows you to connect your Raspberry Pi to your home router via cable. You will need to get your Raspberry Pi connected to your network so that you can access the Internet and install software onto the Raspberry Pi. It can be more convenient to cut out the cable and use a USB WiFi adapter. Such an adapter only costs a few dollars and can be plugged into a USB port. You do need to pick an adapter that is compatible, though. You can find a list of compatible devices at *http://elinux.org/RPi_VerifiedPeripherals*.

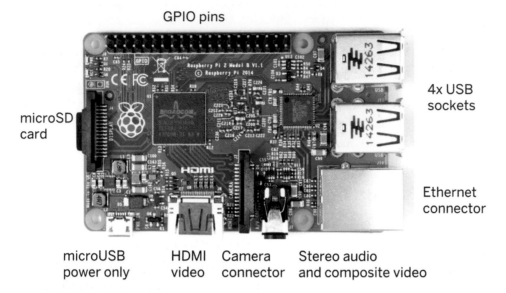

GPIO pins

4x USB sockets

microSD card

Ethernet connector

microUSB power only

HDMI video

Camera connector

Stereo audio and composite video

Figure 3-1 *A Raspberry Pi 2 Model B*

Working our way clockwise around the Raspberry Pi, we come to the stereo audio and composite video connector. This socket will mostly have headphones or an aux lead connected to powered speakers plugged into it, but the connector does also include an extra connection to allow you to connect composite video monitors using a special lead. Generally, the HDMI video connector is more likely to be used to connect a monitor or TV as it is much better quality than composite video. Between the HDMI and audio sockets is a flat cable connector to which a camera specially designed for the Raspberry Pi can be attached.

Next to the HDMI socket is a microUSB connector. This is only used to supply power to the Raspberry Pi using a 5V adapter.

Above the microUSB connector, on the underside of the board, is a slot that takes a microSD card. The Raspberry Pi does not have a conventional hard disk; instead, the operating system and all the files will be contained on a microSD card.

On the top side of the board, you will find a set of header pins. These are called the general-purpose input/output (GPIO) pins; they can be used to connect the Raspberry Pi to various electronic circuits to allow it to control things.

Before the Raspberry Pi B+ was released, Raspberry Pis only had 26 pins on the GPIO connector rather than the 40 shown in Figure 3-1. All the projects in this book use just the 26 original pins that were kept on the new models, so even if you have an old Raspberry Pi, all the projects should still work.

Setting Up Your Raspberry Pi

You'll need to connect a keyboard, mouse, and monitor to your Raspberry Pi in order to set it up. Once set up, you can unplug the keyboard, mouse, and monitor, and instead connect to your Raspberry Pi using Secure Socket Shell (SSH) from another computer. However, until you get to that stage, you will need to keep those things connected to set up your Raspberry Pi.

To set up your Raspberry Pi, you need the following items:

- A USB keyboard and mouse (standard PC peripherals are just fine)

- A monitor or TV with an HDMI input and an HDMI cable

- A 5V microUSB power supply (at least 1 Amp)

- An Ethernet cable to reach to your router, or a USB to WiFi adapter

- A microSD card (4GB OK but 8GB will give you more room for your own files and any programs that you download; choose a microSD card described as class 10, as this will help with performance)

- A second computer and microSD card adapter to set up the SD card (alternatively, you can buy a microSD card with NOOBS—that is, New Out Of the Box Software— preinstalled)

Figure 3-2 shows a typical Raspberry Pi setup.

At least while you are installing the operating system, you should set up your Raspberry Pi where you can connect it directly to your router so that it has an Internet connection. Once the operating system is installed, you can configure a USB WiFi adapter and switch over to wireless if you prefer.

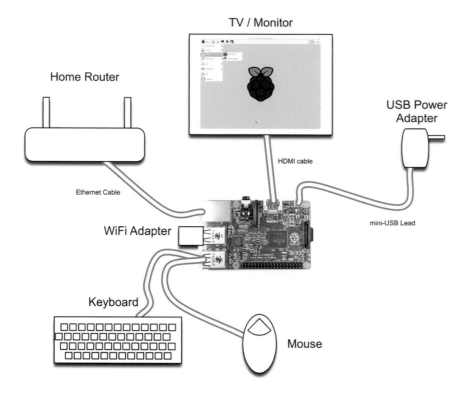

Figure 3-2 *Typical Raspberry Pi setup*

Preparing a MicroSD Card with NOOBS

When the Raspberry Pi was first launched, you had to use special image writing software to set up an SD card. However, NOOBS changed all that—it's as easy to install as copying files onto the SD card and no special formatting is required.

You will find the latest, up-to-date instructions on using NOOBS to set up your microSD card here: *https://www.raspberrypi.org/help/noobs-setup/*.

When your Raspberry Pi boots up into NOOBS it will offer you a choice of operating system (Figure 2-3). Make sure that you select the recommended option Raspbian.

After a lot of file copying and a reboot, your Raspberry Pi will be ready to use. At this point, if you have a USB WiFi adapter, you can configure it and join your wireless network using the WiFi Config utility that you will find under the Preferences section of the desktop menu.

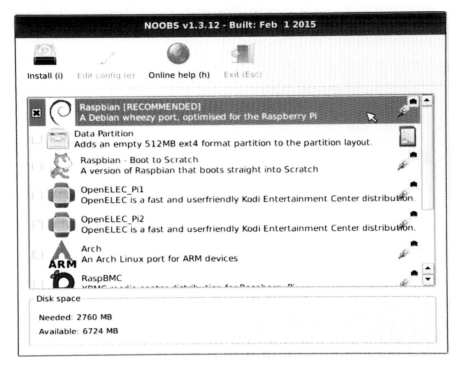

Figure 3-3 *NOOBS offering a selection of operating systems to install*

Setting Up SSH

For many of the projects and experiments in this book, having a keyboard, mouse, and monitor attached to your Raspberry Pi is more of a hindrance than a help. SSH gives you command-line access to your Raspberry Pi over your network from a second computer.

This means that, once you have finished setting it up, the only things that need to be plugged into your Raspberry Pi are a power lead and either a network cable or USB WiFi adapter.

Click the LXTerminal icon at the top of your Raspberry Pi desktop (Figure 3-4). In the LXTerminal window that opens, type the following command (don't include the $—that's the command-line prompt):

```
$ sudo raspi-config
```

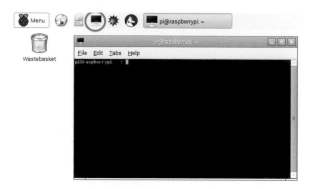

Figure 3-4 *Starting LXTerminal*

This will open the rasp-config tool for setting up your Raspberry Pi. Use the arrow keys to move down to the Advanced menu option and press Enter. Then use the arrow keys again to select SSH (Figure 3-5).

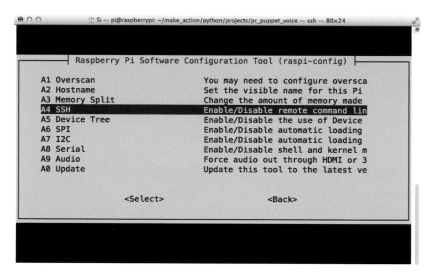

Figure 3-5 *Enabling SSH using raspi-config*

Click Enable, then select the Finish menu option.

Now that your Raspberry Pi is enabled for SSH remote access, it's time to connect from your computer.

If you have a Mac or Linux computer, it will have its own terminal program that you can use to connect to the Raspberry Pi, so skip ahead to "SSH on Mac or Linux" on page 32.

Finding the IP Address of a Raspberry Pi

Before you can connect to your Raspberry Pi using SSH from another computer on your network, you need the Raspberry Pi's IP address.

To find this, run the following command on your Raspberry Pi in the LXTerminal:

```
$ hostname -I
```

This will return a four-part number such as the following (this is your IP address):

```
192.168.1.23
```

SSH on a Windows Computer

If you use Microsoft Windows, you should download and install Putty (*http://www.putty.org/*).

The installation looks a little odd, but actually, you just download *putty.exe,* which is the program itself. You can just save it somewhere (say on your desktop) and run it by double-clicking it. This will open the PuTTY Configuration window (Figure 3-6).

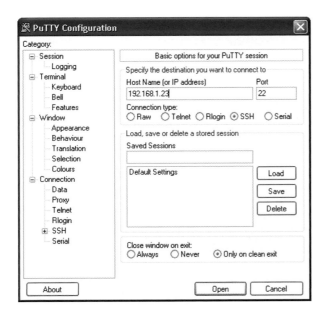

Figure 3-6 *The PuTTY Configuration window*

Enter the IP address of your Raspberry Pi in the "Host Name (or IP address)" field (see "Finding the IP Address of a Raspberry Pi" on page 31), and click Open. You will then be prompted to log in to your Pi (Figure 3-7). Enter a username of *pi* and a password of *raspberry* and that's it, you are in. You can now type commands into PuTTY on your main computer, and they will be executed on your Raspberry Pi.

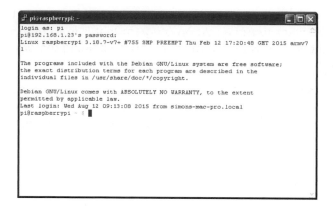

Figure 3-7 *Remote control of your Raspberry Pi with SSH*

SSH on Mac or Linux

If you use a Mac or a Linux computer, the software you need to connect to your Raspberry Pi is already on your computer. Open a terminal session and type the following command, substituting the IP address of your Raspberry Pi in place of the IP address of my Raspberry Pi (192.168.1.23):

```
$ ssh 192.168.1.23 -l pi
```

The first time you do this, you will see the following message:

```
The authenticity of host '192.168.1.23 (192.168.1.23)' can't be established.
RSA key fingerprint is 48:8f:c3:07:c2:04:9e:8b:59:ed:53:2b:0b:d0:aa:e5.
Are you sure you want to continue connecting (yes/no)? yes
Warning: Permanently added '192.168.1.23' (RSA) to the list of known hosts.
pi@192.168.1.23's password:
```

You can go ahead and confirm the authenticity of the computer you are connecting to by typing *yes*. From now on, any commands that you type are actually happening on the Raspberry Pi rather than your other computer.

The Linux Command Line

If you are a Microsoft Windows or Mac user, you may never have needed to use the command line to interact with your computer. Linux, which runs on the Raspberry Pi, does, however, need you to type commands into a command line when installing software, copying and renaming files, running programs, or editing files.

You have already used the Raspberry Pi command line with the LXTerminal to set up SSH on your Raspberry Pi; now you can execute commands on the Raspberry Pi over SSH or directly using LXTerminal.

You may have noticed that whenever LXTerminal or your SSH session is ready for you to type a command, the last character on the line is the $ character. This is called the dollar prompt and is Linux's way of telling you that it's ready for the next command.

The Linux command line has the concept of the *current* directory. This is the directory that you are currently working in. So, if you want to run a Python program that is in a certain directory, you will normally need to set your directory to the directory containing the program before you run it. The command to do this is called cd (*change directory*).

When you first start an LXTerminal session the current directory is */home/pi*. This is called your home directory. If your home directory contains a directory called *make_action*, then you can change to that directory by typing the following (which is changing directory relative to your current directory):

```
$ cd make_action
```

You can also type the whole path for the directory like this:

```
$ cd /home/pi/make_action
```

The Raspberry Pi code in this book is all written in the Python programming language. To run a Python program called *test.py* you would use this command:

```
$ python test.py
```

Another command that you will encounter quite often is sudo (*subsititute user do*). This is used to run the command that follows it as if you were a superuser. This is necessary because Linux tries to protect you from accidentally trashing your own system; superuser access is required to delete important files and to perform other important actions. Superuser access is also required to access the GPIO pins, which you will need to do a lot in this book.

For example, if *test.py* uses the GPIO pins, you would need the following command to run it:

```
$ sudo python test.py
```

Although all the Python code for this book is available as a download (see "The Book Code" on page 34), there will be times when you want to modify a program or other file. You can do this using the nano editor, which will also work just fine over SSH. To edit a file, just enter the command nano followed by the file you want to edit (if no file of that name exists, then nano will create it when you save the edit):

```
$ nano test.py
```

Figure 3-8 shows the nano editor in action.

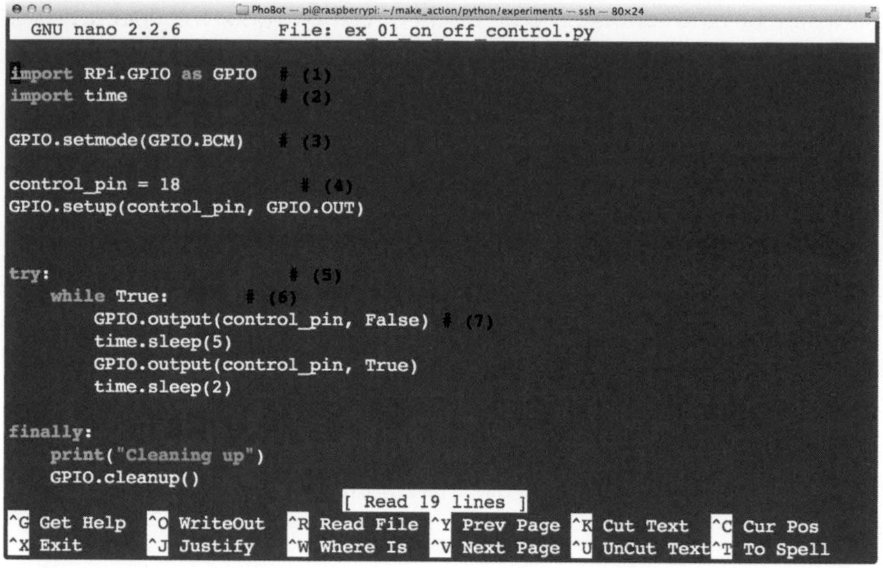

```
800                    PhoBot — pi@raspberrypi: ~/make_action/python/experiments — ssh — 80×24
  GNU nano 2.2.6              File: ex_01_on_off_control.py

import RPi.GPIO as GPIO     # (1)
import time                 # (2)

GPIO.setmode(GPIO.BCM)      # (3)

control_pin = 18            # (4)
GPIO.setup(control_pin, GPIO.OUT)


try:                        # (5)
    while True:             # (6)
        GPIO.output(control_pin, False) # (7)
        time.sleep(5)
        GPIO.output(control_pin, True)
        time.sleep(2)

finally:
    print("Cleaning up")
    GPIO.cleanup()
                      [ Read 19 lines ]
^G Get Help   ^O WriteOut   ^R Read File  ^Y Prev Page  ^K Cut Text   ^C Cur Pos
^X Exit       ^J Justify    ^W Where Is   ^V Next Page  ^U UnCut Text ^T To Spell
```

Figure 3-8 *The nano editor*

Because nano is designed to work in a command-line environment such as the LXTerminal or over SSH, you need to use the cursor keys to move around the file rather than the mouse. When you are ready to save the file, press Ctrl-X and then Y followed by Enter to confirm the save.

The Book Code

The Raspberry Pi does not use a separate programming computer, so you just need to get all the program files used in the book onto the Raspberry Pi itself.

Although you can fetch the files down as a ZIP archive, as you would for the Arduino code (described in Chapter 2) a better way is to use Git, which is a tool for managing program code.

Git keeps track of changes made to the code and will even merge together code from different programmers. The popular public source code repository GitHub uses Git. Git is pre-installed on Raspbian, so all you need to do on your Raspberry Pi to fetch all the code is to open an LXTerminal session, and run the following command:

```
$ git clone https://github.com/simonmonk/make_action.git
```

This will create a directory called *make_action*. In here you will find a *python* directory that contains two subdirectories named *experiments* and *projects*.

The advantage of using Git to fetch the program code is that any time you want to see if there are minor fixes or improvements to the book's code, you can get these changes just by running:

```
$ git pull
```

Programming Guide

The best way to learn how to program is to start by modifying some existing simple programs and gradually understand how things work. All the programs that you need for this book are available for download, so programming expertise is not essential. However, it's nice to have a basic idea of how things work.

Hello, World

Traditionally, when you're working with a new language, the first program you learn to write just prints out the words "Hello, World" on the screen. To try this out, fire up nano using the following command:

```
$ nano hello.py
```

The *.py* file extension identifies the program file as being Python. Next, type the following text into the nano editor and then save the file:

```
print('Hello, World')
```

Then run the program:

```
$ python hello.py
Hello, World
```

Python 2 Versus Python 3

Python users have a hard time letting go of an old version of Python. Although the latest version of Python is Python 3, many people stick to Python 2.

Raspbian incudes both Python 2 and Python 3. If you want to run your "Hello, World" program using Python 3, you would use:

```
$ python3 hello.py
```

The result would be just the same.

So, why do so many Python users stick to version 2 and why does this book use version 2?

The answer is that not all Python libraries have been successfully ported to Python 3. As a case in point, the serial interface library for I2C that you will use in Chapter 14 does not work with Python 3 at the time of writing this book.

Tabs and Indents

Programmers have a style of laying out their code to make it easier to understand. In Arduino C, blocks of code inside a function or `if` statement are indented so you can see which lines of code belong to which command or function. In Python, this is not a matter of personal preference—Python insists upon it.

In Python, there are no { or } to show the start and end of a block of code—you have to use indentation to show which lines belong together.

For example, take a look at this fragment of code:

```
while True :
    GPIO.output(control_pin, False)
    time.sleep(5)
    GPIO.output(control_pin, True)
    time.sleep(2)
print("Finished")
```

This is a while loop (just like while in Arduino C, as described in "Loops" on page 21). In this case, the condition is True. The value True is always True, the loop will continue forever. At the end of the first line, there is a colon (:). The colon indicates that what follows will be a block of code. So the colon in Python is just like the { in Arduino C, but there is no closing marker.

The lines inside the block must be tabbed in from the line above. As long as you are consistent, it does not matter how many spaces (or tabs) you indent the code. Most programmers tend to use four space characters for each level of indentation.

At the end of the block, there is no marker, but instead you just stop indenting the code. So in the preceding example, the final print command is outside of the loop. Because the while loop continues forever, this last line will never actually get run.

Variables

Variables in Python are like Arduino C, but differ because when you first use a variable, you do not have to specify whether it's an int or a float or something else. You just assign it a value and there is nothing (apart from good sense) to stop you assigning the same variable to be an integer value one minute and a string the next. For example, the following code is perfectly legal, but not very sensible:

```
a = 123.45
a = "message"
```

Incidentally, when using strings, you can either use double quotes (as in the preceding example) or single quotes.

if, while, etc.

Python has an if and else structure just like Arduino C, but it uses a colon and indentation to indicate the blocks of code inside the if statement:

```
if x > 10 :
    print("x is big!")
else:
    print("x is small")
```

As you work your way through the book, you will learn more about loops and other programming structures.

The RPi.GPIO Library

Like Arduino C, Python supports libraries—you just need to import them before you can use them in a program.

This book is going to be doing a lot with the GPIO connector on the Raspberry Pi. The most popular Python library to control GPIO pins and the one used in this book is RPi.GPIO.

This library is preinstalled with Raspbian, so you do not need to install anything; you can just import it into your program and start using it.

Confusingly, there are two naming conventions for the pins on a Raspberry Pi's GPIO connector: one uses the positions of the pins on the connector (1, 2, 3, 4, etc.) and the other uses the functions of the pins. In the early days of Raspberry Pi, both were used fairly frequently. These days, pretty much everyone refers to pins based on their function. The RPi.GPIO library will work with both ways of pin naming, but you have to tell it which to use, so you will see the following code near the top of every Python program in this book:

```
import RPi.GPIO as GPIO
GPIO.setmode(GPIO.BCM)
```

The first line imports the RPi.GPIO library and the second specifies the Broadcom (BMC), which is the name of the chip at the center of the Raspberry Pi.

Unlike the Arduino, the Raspberry Pi does not have analog inputs. The RPi.GPIO library does, however, support PWM analog outputs.

The GPIO Header

Figure 3-9 shows the connections available on the GPIO header. If you have an older Raspberry Pi, with only 26 pins, then the pins below the dotted line will not be present on your board. The projects in this book can be completed on all Raspberry Pi models, so only those 26 pins will be used.

This diagram is repeated at the end of the book in Appendix B as a handy reference. Unlike the Arduino, a Raspberry Pi does not have the pin names printed on the printed circuit board (PCB), making it difficult to tell which pin is which. You can make this easier by buying or making a template to fit over the pins, such as the Raspberry Leaf (*https:// www.adafruit.com/products/2196*).

Notice how some pins have a number (say, 2) but also a second function listed next to them (say, SDA). When using the GPIO pins in your code, you always refer to them just by number.

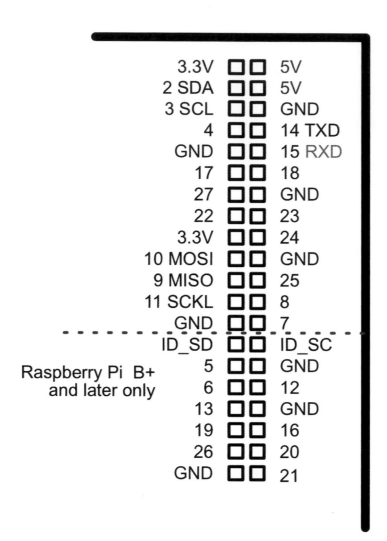

Figure 3-9 *The GPIO header*

Digital Outputs

This example shows how you would set GPIO pin 18 (as indicated in Figure 3-9) to be a digital output and then set the output high:

```
import RPi.GPIO as GPIO
GPIO.setmode(GPIO.BCM)
GPIO.setup(18, GPIO.OUT)
GPIO.output(18, True)
```

When setting an output using `GPIO.output`, you can set it high using a value of `True` or `1`, or low using `False` or `0`.

Digital Inputs

Digital inputs on a Raspberry Pi work in much the same way as digital inputs on an Arduino:

```
GPIO.setup(18, GPIO.IN)
value = GPIO.input(18)
print(value)
```

You can also specify that the input has a pull-up resistor on it (see "Digital Inputs" on page 17) like this:

```
GPIO.setup(switch_pin, GPIO.IN, pull_up_down=GPIO.PUD_UP)
```

Analog Outputs

Using analog outputs on a Raspberry Pi is a two-stage process. You first have to set up the pin to be an output, and then you have to define a PWM channel that uses that output.

You will find a fully explained example that uses this in "Experiment: Mixing Colors" on page 86.

Summary

In this chapter, you have learned how to set up your Raspberry Pi and get to grips with the Linux command line, and learned some of the basics of Python language that you will need to control things using the GPIO pins.

Now it's time to get started with some practical experience. Chapter 4 will get you started using your Arduino and Raspberry Pi for real.

Quickstart

4.

It's always better—and definitely more fun—to get some practical experience rather than just read about things. This chapter is designed to get you up and running using a solderless breadboard with some electronic components. You'll learn how to control an LED and then a motor from your Arduino or Raspberry Pi.

Solderless Breadboard

In most cases, when connecting external electronics to a Raspberry Pi or Arduino, it is not possible to connect the device (say, a motor) directly to the Raspberry Pi or Arduino. You will need to connect some extra electronic components to make it all happen. In any case, even if you are just lighting an LED, you will need some way of attaching the LED to your Raspberry Pi or Arduino.

A great way of connecting things up without the need for soldering is to use a *solderless breadboard*. The breadboard (as it's usually called) was invented as a tool for electronics engineers to prototype their designs before committing the projects to a more permanent soldered form. Breadboards allow you to experiment with electronics and make your own projects without the need for any soldering.

Figure 4-1 shows a breadboard populated with the components from "Experiment: Controlling a Motor" on page 53.

You can see how the breadboard is useful both to hold and connect components and wires.

The breadboard used throughout this book is often referred to as a half-sized breadboard or 400-point breadboard (because it has 400 holes in it). There are bigger and smaller breadboards available, but this size is about right for all the projects and experiments in this book.

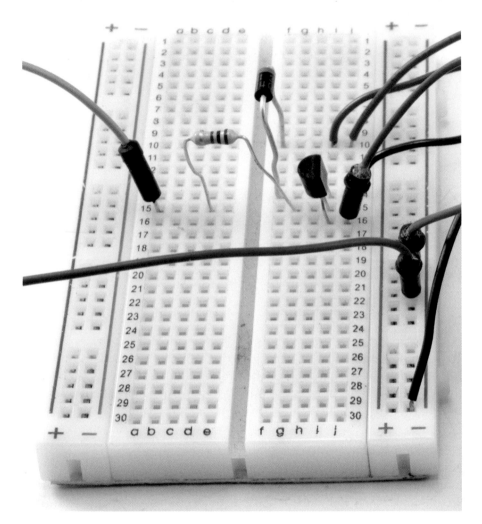

Figure 4-1 *Using a solderless breadboard*

This type of breadboard has two columns on each side of the breadboard. These are usually marked with red and blue lines. For any one of these columns, all the holes in that column are connected together electrically behind the plastic of the board. Although these columns can be used for anything, they are most often used for the positive and negative supply to your circuit.

The main body of the board is split into two banks, or rows, of five holes each down the whole length of the board. Each of these rows of five holes has all five holes connected together by a clip behind the plastic. To connect the leg of one component to another, both leads just need to be pushed into holes on the same row of the breadboard.

How a Breadboard Works

Behind the holes in the front plastic face of a breadboard, there are metal clips designed to grip wires and component leads.

Figure 4-2 shows a dismantled breadboard with one of the clips removed, so that you can see how it works behind the plastic. I would not recommend taking your breadboard apart like this, as generally it will never quite be the same again once you have put it back together.

Figure 4-2 *The inner workings of a breadboard, with one clip removed*

Connecting a Breadboard to the Arduino

The Arduino GPIO connections (confusingly often called pins) are really sockets. To connect one of these pins to one of the rows of a breadboard, you can use a "male-to-male" jumper wire, as shown in Figure 4-3.

Figure 4-3 *Connecting the Arduino to a breadboard*

These jumper wires have a flexible wire with a little plug on each end. It's a good idea to keep a good stock of such wires in different lengths ready to connect things up.

You can buy jumper wire packs (see Appendix A) in a variety of sizes and colors. It is well worth getting yourself a little kit of breadboard and assorted jumper wires. Different colors make it easier to see how the wires are connected up, especially when you have a whole load of wires on the breadboard.

Connecting a Breadboard to the Raspberry Pi

The GPIO connection pins of a Raspberry Pi are actually pins rather than sockets. This means that you cannot use the male-to-male jumper wires that you did for the Arduino. Instead, you need to use (you guessed it) female-to-male jumper wires. These have a socket on one end that fits over a pin on the Raspberry Pi's GPIO header, and a pin on the other end of the lead that fits into the breadboard.

Figure 4-4 shows how you can make connections from the GPIO pins of a Raspberry Pi to a particular row of your breadboard using female-to-male jumper wires.

Figure 4-4 *Connecting a Raspberry Pi*

 Identifying GPIO Pins on a Raspberry Pi

The GPIO pins on a Raspberry Pi are not labeled on the board, so to save you from the trouble of carefully counting pins, you can use a GPIO template to fit over the GPIO connector. The one shown in Figure 4-4 is a Raspberry Leaf, available from Adafruit and MonkMakes.com. Other templates are also available.

Downloading the Software

All the software for this book, both Arduino sketches (as Arduino programs are called) and Python programs for the Raspberry Pi, are available from the book's GitHub repository (*https://github.com/simonmonk/make_action*).

You can find instructions for uploading the Arduino sketches from your everyday computer to your Arduino in "The Book Code" on page 14 in Chapter 2.

To get the Python programs for the book onto your Raspberry Pi, follow the instructions in "The Book Code" on page 34 in Chapter 3.

Experiment: Controlling an LED

In the Arduino world, it has become traditional to, as a first exercise, make an LED blink on and off. In this first experiment, you will also make an LED blink, first using an Arduino and then with a Raspberry Pi.

This is a nice, easy project to get you started. There are just two components to fit onto the breadboard: an LED and a resistor. All LEDs need a resistor to limit the current flowing through them. You will find more in-depth information about this in Chapter 6.

Parts List

Whether you are using a Raspberry Pi or Arduino (or both), you are going to need the following parts to carry out this experiment:

Part	Source
Red LED	Adafruit: 297
	Sparkfun: COM-09590
470Ω 1/4W resistor	Mouser: 291-470-RC
400-point solderless breadboard	Adafruit: 64
Male-to-male jumper wires (Arduino only)	Adafruit: 758
Female-to-male jumper wires (Pi only)	Adafruit: 826

This table lists the supplier and product code for each part. You will also find more information about all the parts used in this book in Appendix A.

Breadboard Layout

The breadboard layout for this project is shown in Figure 4-5. This is the same whether you are using an Arduino or a Raspberry Pi, but the way that you link the breadboard to the Raspberry Pi or Arduino will be different.

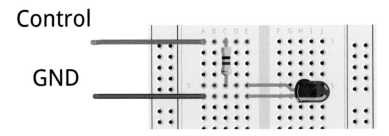

Figure 4-5 *The breadboard layout for controlling an LED*

Figure 4-5 shows how the current from an Arduino or Raspberry Pi output pin will flow first through the resistor and then the LED to make it light.

It does not matter which way around the resistor goes, but the LED must have the positive lead toward the top of the breadboard. The positive lead of an LED is slightly longer than the negative lead. Also, the edge of the LED next to the negative lead will have a flat edge.

Resistors will be covered in more depth in Chapter 5.

Arduino Connections

Connect the Arduino sockets GND and D9, as shown in Figure 4-6. You can see the actual Arduino and breadboard in Figure 4-7.

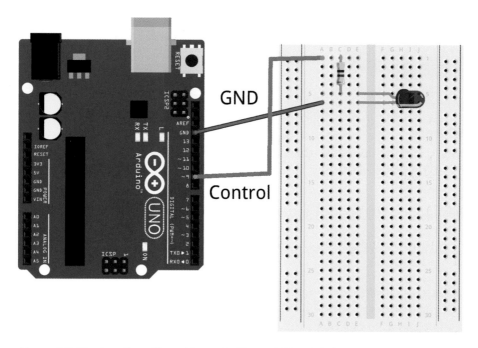

Figure 4-6 *The breadboard layout for controlling an LED with Arduino*

Figure 4-7 *Using a breadboard with Arduino*

Arduino Software

The Arduino sketch can be found in the *arduino/experiments/on_off_control* directory, which you'll find in the place where you downloaded the book's code (see "Downloading the Software" on page 45).

The program will turn the LED on for 5 seconds, then off for 2 seconds, and then repeat. Here is the full code:

```
const int controlPin = 9; // ❶

void setup() {  // ❷
  pinMode(controlPin, OUTPUT);
}

void loop() {  // ❸
  digitalWrite(controlPin, HIGH);
  delay(5000);
  digitalWrite(controlPin, LOW);
  delay(2000);
}
```

❶ The first line defines a constant called `controlPin` as pin 9. Although the Arduino Uno has digital pins 0 to 13 and analog pins 0 to 5, the convention is that if the Arduino code refers to just a number by itself (in this case, 9) then that refers to the digital pin. If you wish to refer to one of the six analog pins, you must prefix the pin number with the letter A.

❷ The `setup()` function specifies the pin as being a digital output using `pinMode`.

❸ The loop() function, which will repeat indefinitely, first sets the controlPin (9) high (5V) to turn the motor on. It then delays for 5,000 milliseconds (5 seconds) and then sets the controlPin low (0V) to turn off the motor. The next line then delays for another 2 seconds before the loop starts again.

Arduino Experimentation

Upload the program to your Arduino. As soon as the Arduino restarts as part of the upload process, it will be running your code and the LED should start blinking.

If the LED does not blink, then check all the connections and that the LED is the right way around (longer LED lead to the top of the breadboard).

Try altering the numbers in the delay functions to alter how long the LED stays on on for each cycle. You will need to upload the program again each time you make a change to it.

Raspberry Pi Connections

Unlike the Arduino, a Raspberry Pi does not have any labels next to its GPIO pins to tell you which is which. This leaves you with two options: you can use a diagram of the GPIO pinout (see Appendix B) and count down to find the pin you need, or you can use a pin identification template that fits over the GPIO pins, such as the Raspberry Leaf shown in Figure 4-8. Figure 4-9 shows the breadboard layout and connections to the Raspberry Pi.

Figure 4-8 *Connecting a breadboard to a Raspberry Pi*

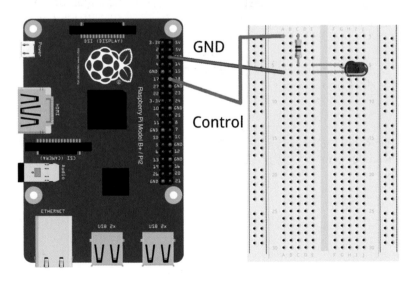

Figure 4-9 *The breadboard layout for controlling an LED with a Raspberry Pi*

Raspberry Pi Software

You do not need a separate computer to program the Raspberry Pi—it's possible to write and run the program on the Raspberry Pi itself. To do so, use the following program (you can find it in the *on_off_control.py* file in the *python/experiments* directory):

```
import RPi.GPIO as GPIO     # ❶
import time                 # ❷

GPIO.setmode(GPIO.BCM)      # ❸

control_pin = 18            # ❹
GPIO.setup(control_pin, GPIO.OUT)


try:                        # ❺
    while True:             # ❻
        GPIO.output(control_pin, False) # ❼
        time.sleep(5)
        GPIO.output(control_pin, True)
        time.sleep(2)

finally:
    print("Cleaning up")
    GPIO.cleanup()
```

The program is quite similar to its Arduino counterpart:

❶ To gain access to the GPIO pins of the Raspberry Pi, there is a Python library called RPi.GPIO created by Raspberry Pi enthusiast Ben Croston. The first line of code imports this library so that it can be used in your program. The RPi.GPIO library comes preinstalled in all recent versions of the standard Raspbian distribution, so you shouldn't need to install it, unless you're using an old version of Raspbian. In that case, the easiest way to install it is to update your system, which you should probably do anyway, by issuing the following command in a terminal session:

```
$ sudo apt-get upgrade
```

❷ The time library also needs to be imported, as this is used to cause the delays between turning the LED on and off.

❸ The GPIO.setmode(GPIO.BCM) line must be included in every Python program that you write that controls GPIO pins, before you set the mode of the pins or use them in any way. The command tells the GPIO library that the pins are to be identified by their Broadcom (BCM) names rather than by the position of the pins. The RPi.GPIO library supports both naming schemes, but the Broadcom naming convention is the more popular of the two, and it's the one used throughout this book.

There are no separate setup() and loop() functions used when programming Arduino; instead the things that would go in setup() just appear near the start of the program and a `while` loop that continues forever takes care of what would normally be in `loop()` in an Arduino.

❹ The variable `control_pin` identifies GPIO pin 18 as the pin that you are using to control the LED and then this pin is defined as being an output using `GPIO.setup`.

❺ Now you come to the equivalent of the Arduino's loop function. This is contained in a `try/finally` construction. The point of this is that if any error occurs in the program, or you just stop the program by pressing Ctrl-C in the terminal window where you ran the program, then the cleanup code in the finally block will be run.

You could omit this code and just have the `while` loop by itself, but the `cleanup` code automatically sets all the GPIO pins back to a safe state as inputs, making it much less likely that an accidental short or wiring error on your breadboard could damage your Raspberry Pi.

❻ The `while` loop has a condition of `True`. This may seem strange, but it's just the way you make some code continue indefinitely in Python. In fact, the program will just keep looping around the commands inside the `while` loop until you kill the program with Ctrl-C or unplug your Raspberry Pi.

❼ Inside the loop, the code is very similar to the Arduino counterpart. The GPIO pin is set to `True` (high) then there is a delay of 5 seconds, then the GPIO pin is set to `False` (low) and another 2-second delay happens before the cycle starts again.

Raspberry Pi Experimentation

To access the GPIO pins, Linux superuser privileges are required. To run your program, change into the directory containing *on_off_control.py* and then run the program using this command:

```
$ sudo python on_off_control.py
```

Including `sudo` at the beginning of the command allows it to be run as a superuser. When you have had enough of the motor running, press Ctrl-C to quit the program.

Comparing the Code

The overall structure of both Arduino C and Python code is quite similar, but the code itself is different in style. Also, when naming variables or functions, C uses the convention (that is, each word after the first word is capitalized), whereas Python uses snake_case (that is, words are separated by underscores).

Table 4-1 compares some of the key differences between the two programs.

Table 4-1 *Arduino C and Python*

Command	Arduino C code	Python code
Define a constant for a pin	`const int controlPin = 9;`	`control_pin = 18`
Set a pin to be an output	`pinMode(controlPin, OUTPUT)`	`GPIO.setup(con trol_pin, GPIO.OUT)`
Set an output high	`digitalWrite(controlPin, HIGH);`	`GPIO.output(con trol_pin, True)`
Set an output low	`digitalWrite(controlPin, LOW);`	`GPIO.output(con trol_pin, False)`
Delay for 1 second	`delay(1000);`	`time.sleep(1);`

Experiment: Controlling a Motor

Now that you can use the Raspberry Pi and Arduino to turn an LED on and off, let's apply that knowledge so that you can use them to turn a motor on and off. This will use the exact same software as "Experiment: Controlling an LED" on page 46, but you will use a transistor to switch the DC motor.

You will learn a lot more about DC motors in Chapter 7. DC motors are the kind of small motor that you might find in a hand-held fan or a toy car. They are probably the easiest type of motor to use—you supply them with a voltage across their two terminals and their shaft rotates.

Because nearly all motors require too much current to be driven directly from the digital output of the Raspberry Pi or Arduino, a transistor is used to allow a small current from the Raspberry Pi or Arduino to control a much bigger current to the motor.

The same electronics hardware is used for both the Raspberry Pi and the Arduino, and in both cases is built on the same solderless breadboard.

This is a quickstart chapter, and as such, some of what is going on in this experiment will not be explained until later chapters. The layout also has a lot more components than the first experiment, so take care to ensure that all the component leads are in the right holes and that components that need to be the right way around are correct.

Parts List

Whether you are using a Raspberry Pi or Arduino (or both), you are going to need the following parts to carry out this experiment:

MPSA14 Darlington transistor	Mouser: 833-MPSA14-AP
1N4001 diode	Adafruit: 755
	Sparkfun: COM-08589
	Mouser: 512-1N4001
470Ω 1/4W resistor	Mouser: 291-470-RC
Small 6V DC motor	Adafruit: 711
6V (4 x AA) battery box	Adafruit: 830
400-point solderless breadboard	Adafruit: 64
Male-to-male jumper wires	Adafruit: 758
Female-to-male jumper wires (Raspberry Pi only)	Adafruit: 826

If you are planning to try this experiment with a Raspberry Pi, you will need female-to-male jumper wires to connect the Raspberry Pi GPIO pins to the breadboard.

Breadboard Layout

The breadboard layout for this project is shown in Figure 4-10.

When you put the components onto the breadboard, you need to make sure that the transistor is positioned correctly—the flat side with the writing on it should be on the right-hand side. In addition, you need to check the placement of the diode—it has a stripe at one end that should face the top of the board.

Because playing with something is often more fun than understanding the details of how it works, we won't fully discuss how this experiment works until Chapter 5.

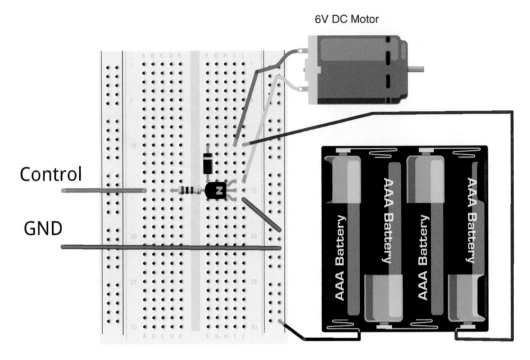

6V DC Motor

Control

GND

Figure 4-10 *The breadboard layout for controlling a motor*

Experimenting Without Arduino or Raspberry Pi

Before you start connecting the breadboard to an Arduino or Raspberry Pi, you can experiment with it and test it out.

The transistor is acting as a switch (more on that in Chapter 5). Eventually, two wires will be connected to the Arduino or Raspberry Pi, GND and Control. The *GND (ground) connection* represents zero volts for both the breadboard circuit and the Arduino and Raspberry Pi. The *Control connection* will turn the motor on if it is connected to any voltage over about 2V and the motor will be off if that voltage is less than that.

You can try that out using a male-to-male jumper wire before getting an Arduino or Raspberry Pi involved. Attach one end to the same row as the left lead of the resistor and touch the other end to the top lead of the diode which is also connected to battery + (Figure 4-11). When you do this, the motor will start, and when you take the header lead away from the diode lead, the motor should stop again.

Figure 4-11 *Testing the circuit before connecting to an Arduino or Raspberry Pi*

Arduino Connections

Now that you are sure that the control lead from the breadboard will indeed turn the motor on and off, you can connect it to one of Arduino GPIO pins using a male-to-male jumper wire. Use the connection labeled "9" on the Arduino, as shown inFigure 4-12. Note this is the same control pin that you used for the LED in "Experiment: Controlling an LED" on page 46.

You will also need to connect the other GND lead to the Arduino GND leads. This is also shown in Figure 4-12.

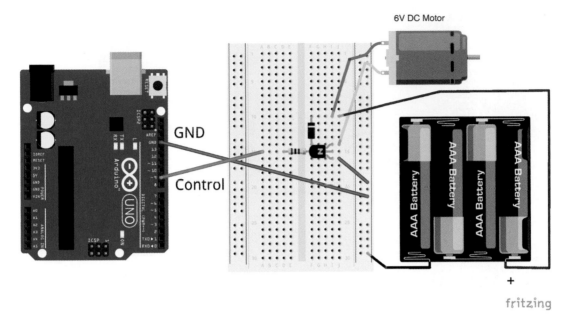

Figure 4-12 *The breadboard layout for controlling a motor with an Arduino*

Arduino Experimentation

If you still have the program on your Arduino from "Experiment: Controlling an LED" on page 46, you do not need to upload anything. If you no longer have that program on your Arduino, follow the instructions in that section so that you can upload it again.

As you did with the LED, try altering the numbers in the delay functions to alter how long the motor remains on for each cycle.

Raspberry Pi Connections

You can now disconnect the breadboard from the Arduino and connect it to the Raspberry Pi. Connect GPIO pin 18 (the control pin) and GND (ground), as shown in Figure 4-13.

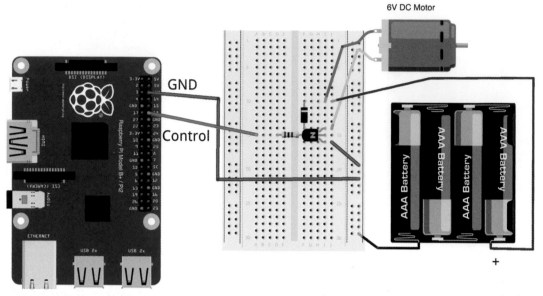

GND

Control

6V DC Motor

fritzing

Figure 4-13 *The breadboard layout for controlling an LED with a Raspberry Pi*

Raspberry Pi Experimentation

The Raspberry Pi program for this experiment is also the same as for controlling the LED.

Run your program, change directory to the directory containing *on_off_control.py*, and then run the program using this command:

```
$ sudo python on_off_control.py
```

When you have had enough of the motor running, press Ctrl-C to quit the program.

Summary

The goal of this chapter was to get you going with actually making something. The next chapter more fully describes the theory behind how this project works. We'll also discuss how to select the right transistor for the job and identify components.

Basic Electronics   5

In "Experiment: Controlling an LED" on page 46, you were launched right into controlling a motor using a transistor, without any real explanation as to how this worked. If you already know a little about electronics, you might well be familiar with transistors and be able to skip much of this chapter. If, on the other hand, you are new to electronics, you will find this explanation helpful.

Current, Voltage, and Resistance

In this section, you will need to refer to the schematic diagram for "Experiment: Controlling a Motor" on page 53 that you built on the breadboard. This is shown in Figure 5-1.

The schematic view is just a different and more abstract way of representing what you made on the breadboard. Instead of diagramming how components look in real life, they are represented in a way that indicates what they do.

Current

The zig-zag line of resistor R1 gives a visual clue that this component will restrict the flow of current. In electronics, the word *current* refers to a flow of electrons through wires or components. You can think of it as electrons flowing from one place on your circuit to another. As an example, current is flowing out of a GPIO output on an Arduino or Raspberry Pi and into resistor R1. The current flows out of R1 and into the center connection (the base) of transistor Q1.

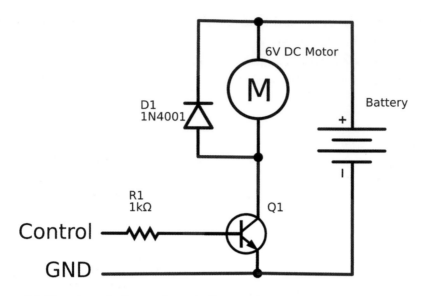

Figure 5-1 *The schematic diagram for controlling a motor*

A small current flowing into the base of a transistor allows a much bigger current to flow through the two righthand connections of the transistor, the collector (top) and emitter (bottom). That is how the tiny current from the GPIO pin can control the much bigger current needed for something like a motor. It can be helpful to think of it as a digital switch that can be turned on and off with a small current.

The unit of measurement used for current is the *ampere*, or more commonly *amp*, which is further abbreviated to just *A*. A current of 1A is actually a relatively high current when dealing with an Arduino or Raspberry Pi, so the unit of milliamp (mA) is often used. 1mA is 1/1000 of an amp.

The reason we use the resistor R1 to restrict the flow of current is that the GPIO pins of a Raspberry Pi or Arduino cannot provide enough current to drive a motor directly. In fact, if you try to do that, there is a good chance you will either damage or destroy your Pi or Arduino. A Raspberry Pi can safely deliver about 16mA of current and an Arduino about 40mA.

Voltage

In the same way that water always flows from higher places to lower places, electrical current always flows from parts of the circuit that are at a higher voltage to places that are at a lower voltage. So, in the case of the control GPIO pin in Figure 5-1, when the pin is at 0V, no current will flow out of it and through the resistor and then transistor, to ground, because the GPIO pin and ground are at the same voltage (0V). However, when the pin is high, at 3.3V on a Raspberry Pi or 5V on an Arduino, the current will flow through the resistor and transistor and then to ground.

An important point about schematics and voltage is that all points on a schematic that are connected by a line are the same voltage.

The unit of measurement for voltage is the volt, abbreviated to V. A Raspberry Pi's GPIO pins used as an output are either 3.3V (high) or 0V (low) and those of an Arduino are 5V or 0V.

Ground

The line at the bottom of Figure 5-1 is labeled as ground. You will often see this on schematic diagrams labeled as GND. Ground represents zero volts in a schematic and is the base voltage against which other voltages in the circuit are measured. For example, the top positive terminal of the battery in Figure 5-1 will be referred to as being at 6V because it is 6V higher than ground.

When it comes to connecting different parts of a project together, the grounds from each part will all be connected together. In this case, if you were going to attach this motor control module to an Arduino or Raspberry Pi, then the ground of this motor controller would be connected to one of the ground pins (GND) of the Arduino or Raspberry Pi.

Resistance

Resistors have a value of resistance that is measured in ohms, which is abbreviated to the Greek letter omega (Ω). Resistors span quite a large range of values, so you will find resistor values in the kΩ range (thousands of Ω) and sometimes in the MΩ range (millions of Ω).

The resistor used in "Experiment: Controlling a Motor" on page 53 is a 1kΩ resistor, and you can work out how much this 1kΩ resistor will limit the current using something called *Ohm's Law*. The law states that the current flowing through a resistor will be the difference in voltage across the resistor (in volts) divided by the value of the resistor in ohms. In the case of a Raspberry Pi, the biggest voltage drop that there could possibly be between the GPIO pin and GND is when the GPIO pin is high at 3.3V. So the maximum current that could possibly flow is 3.3V / 1000Ω = 3.3mA.

Raspberry Pi GPIO Pin Current

The maximum current that can be handled by a GPIO pin of a Raspberry Pi is a matter of some debate and uncertainty. I generally use the original figure specified by the makers of the Raspberry Pi (that is, 3mA per GPIO pin). This figure of 3mA was because the original Raspberry Pi only had 14 GPIO pins and its 3V voltage regulator could only make 50mA available for the GPIO pins, which equated to 3mA per GPIO pin, if all the GPIO pins were being used.

This no longer applies with newer Raspberry Pi models (A+, B+, and Pi 2), where there are 24 pins available to use as GPIO pins and a 3V regulator that theoretically can supply up to 1A. If you are using one of these new Raspberry Pi models, you should not try to draw 1A / 24 = 41mA per GPIO pin, because the Raspberry Pi's Broadcom system on a chip (SoC) also has a limit per GPIO pin, which is 16mA.

On a Raspberry Pi A+, B+, or Pi 2, you can theoretically use up to 16mA on as many GPIO pins as you need. However, there are other factors that have a bearing on how much current you can safely draw without damaging the Pi, such as the switching frequency and the total GPIO current that the Broadcom chip can cope with.

In summary, you should follow these guidelines:

- On an original Raspberry Pi 1, you can use 16mA per pin up to an overall total current of 40mA.

- On a Raspberry Pi A+, B+, or Pi 2, a safe approach would be to use no more than 16mA per pin, up to a maximum of 100mA for all pins used.

The resistor is protecting the GPIO pin to limit the current that can be drawn from it, and if every digital output from a Raspberry Pi always starts with a 1kΩ resistor, we can rest assured that our Pi will be safe. Quite often, especially when using LEDs, you will be able to use a lower value of resistor because something else (such as the LED) will be using some of the voltage drop, reducing the current.

Power

When a current passes through a resistor, there is a heating effect. The electrical energy is converted into heat energy and power is the amount of this energy that is converted per second. The unit of power is the watt (W), and the heat that a component generates is calculated by multiplying the voltage across it (in volts) by the current passing through it in amps.

Returning to our 1kΩ resistor from experiment 1, if it has 2.2V across it and a current flowing through it of 2.2mA, so it will produce about 4.8mW of power. That's very little power. However, transistors will also generate heat of the voltage across them multiplied by the current flowing through them. If you used a fairly powerful motor, say 800mA for "Experiment: Controlling a Motor" on page 53, the voltage drop from the collector to the emitter of a fully on transistor will be about 1.2V. So the power converted to heat will be 800mA x 1.2V = 960mW. That will make the transistor get quite hot. If the transistor gets too hot, something inside will eventually melt and the transistor will fail. For this reason, just as the Raspberry Pi and Arduino output pins have a maximum current, so do transistors. Physically larger transistors will usually be rated for larger currents, which is one of the factors that you need to consider when selecting a transistor to control an actuator.

The MPSA14 Darlington transistor used in "Experiment: Controlling a Motor" on page 53 has a maximum current of 1A.

Common Components

In the following subsections, we will look at some of the components used in this book and explain how to use them and also how to select them.

Resistors

Resistors are colorful little devices. If you are trying to work out their value, you can just measure their resistance with a multimeter, or you can read their value from the colored stripes.

Each color has a number associated with it, as shown in Table 5-1.

Table 5-1 *Resistor color codes*

Black	0
Brown	1
Red	2
Orange	3
Yellow	4
Green	5
Blue	6
Violet	7
Gray	8
White	9
Gold	1/10
Silver	1/100

Gold and silver, as well as representing the fractions 1/10 and 1/100, are also used to indicate how accurate the resistor is, so gold is +–5% and silver is +–10%.

There will generally be three of these bands together starting at one end of the resistor, a gap, and then a single band at the other end of the resistor. The single band indicates the accuracy of the resistor value.

Figure 5-2 shows the arrangement of the colored bands. The resistor value uses just the three bands. The first band is the first digit, the second the second digit, and the third "multiplier" band is how many zeros to put after the first two digits.

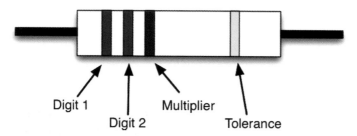

Digit 1   Multiplier

Digit 2   Tolerance

Figure 5-2 *Reading resistor color codes*

So a 270Ω resistor has a first digit 2 (red), second digit 7 (violet), and a multiplier of 1 (brown). Similarly, at 1kΩ resistor will have bands of brown, black, and red (1, 0, 00).

Resistors also have a power rating. Through-hole resistors of the type used in this book are nearly all 1/4W. Other common power ratings are 1/2W, 1W, and 2W, with the resistors becoming increasingly large physically with the power rating.

Transistors

The range of transistors on offer from a component supplier is generally huge and bewildering, so in this book, I have simplified the choice to four transistors that will cover most of the bases, when it comes to switching things on and off.

The transistor in "Experiment: Controlling a Motor" on page 53 is what allows the tiny current of a few milliamps to control the hundreds of milliamps required by the motor. Although transistors can be used in other roles, in this book we will be using them as switches. A small current flowing into the *base* and down to GND through the *emitter* of the transistor will switch a much larger current flowing from the *collector* to the *emitter*.

Figure 5-3 shows a selection of transistors of various types and power handling capabilities.

Figure 5-3 *A selection of transistors*

Transistors are supplied in a fairly small number of standard package types. So when identifying a transistor, you cannot go by appearance; you need to read the name that is printed on it.

The most common packages are TO-92 on the left and TO-220 in the center. Occasionally, for very high-power applications, you may use a transistor with a larger case style, such as the TO-247 package on the right.

The TO-220 and TO-247 packages are both designed to be bolted to heat sinks. It is not necessary to fix these transistors to a heat sink if you are using them at much lower currents than their specified maximum current.

Bipolar transistors

Transistors are made using different technologies that have pros and cons that make them suitable for some situations and unsuitable for others.

The bipolar transistor is where most people start with transistors. These have not changed much since the early days of transistors. They have the advantage that they are very cheap and easy to use for smallish load currents. Their downside is that although a small current passing through the base to emitter of the transistor will result in a bigger current flowing through from the collector to the emitter, the collector current is limited to a multiple of the base current and that multiple (called *gain* or *hFE*) is typically between 50 and 200. So, if a Raspberry Pi is only supplying 2mA to the base, the current flowing through the collector may only be 100mA. This may be a lot less than you were expecting, as the transistor may have a much higher current-carrying capability (say, 500mA) but never get to that limit because the base current is not enough. This is not normally a problem when using an Arduino, as it can supply more current to the base (up to 40mA) by using a lower value resistor in place of the 1kΩ resistor in Figure 5-1. If you selected a resistor value of 150Ω, for example, the base current would increase to I = V / R = (5 – 0.5) / 150 = 30mA. A base current of 30mA would even in the worst case of a transistor with a gain of 50 still result in a collector current of 1.5A.

The voltage calculation in the preceding equation is (5 – 0.5) because the voltage between the base and emitter of a bipolar transistor will be around 0.5V when the transistor is turned on.

In this book, we use just one model of bipolar resistor, the very common 2N3904. Although higher current bipolar transistors are available, there are better transistor technologies to use as the current starts to get higher.

Darlington transistors

When you need a bit more gain, if perhaps you are driving a small motor from a Raspberry Pi that is only capable of supplying a few milliamps into the base, then a Darlington transistor is a good alternative to a regular bipolar transistor and will typically have a gain of at least 10,000.

Darlington transistors are actually made of two bipolar transistors within one transistor package (Figure 5-4). This two-stage arrangement is what gives the Darlington its high gain.

Because there are now two base-emitter junctions, each will drop at least 0.5V when the transistor is on, and so there will be a voltage drop of around 1V rather than the 0.5V of a normal bipolar transistor. In fact, this drop is also applied to the collector voltage and increases with the load current. This means that when controlling a current of 1A, an MPSA14 Darlington transistor may only actually provide 9V when powering a 12V load. Sometimes this matters, but other times it does not.

The transistor used in "Experiment: Controlling a Motor" on page 53 (an MPSA14) is a Darlington transistor. The voltage drop across R1 when using a Raspberry Pi will not actually be 3.3V, but rather 3.3V–1V or 2.2V. So, the current that the Raspberry Pi needs to supply will be 2.2V / 1kΩ = 2.2mA.

As well as the low-power MPSA14, which is good for controlling loads up to 0.5A, I also recommend the higher-power TIP120 Darlington transistor as a standard transistor to keep in your component box.

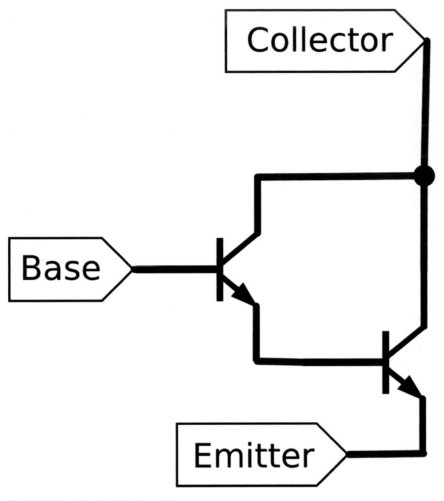

Figure 5-4 *A Darlington transistor*

MOSFETs

Bipolar transistors are essentially current-driven devices—the small base current is amplified into a bigger collector current. There is another type of transistor called the metal-oxide-semiconductor field-effect transistor (MOSFET) that requires very little current to switch, but instead will turn on as long as the voltage at its "gate" connection is greater than a certain threshold.

Figure 5-5 shows the schematic diagram for the transistor; you can see that the symbol implies that the gate is not directly connected to the rest of the transistor.

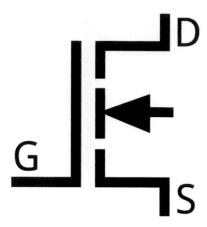

Figure 5-5 *The symbol for a MOSFET*

Note that instead of a base, collector, and emitter, a MOSFET has a gate, drain, and source. Intuitively, you might expect the source and collector to be equivalents, but actually the drain is equivalent to the collector of a bipolar transistor.

Switching using a MOSFET actually makes a lot of sense if you are using an Arduino or Raspberry Pi, because almost no current is needed; you just need to make sure that the voltage at the gate is greater than the gate threshold voltage for the transistor. The gate threshold voltage is the voltage at which the MOSFET turns on, allowing a current to flow from drain to source. Figure 5-6 shows how to connect up a MOSFET to control a load. This is really just the same as for a bipolar transistor.

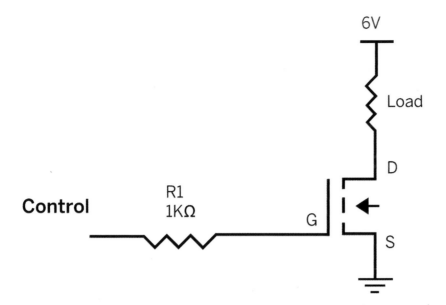

Figure 5-6 *Using a MOSFET*

The schematic diagram shown in Figure 5-6 uses two symbols we have not seen before. Connected to the source of the transistor (S) is a row of three parallel lines of reducing length. This is the symbol for ground, and using it in schematics reduces the number of connecting lines that you need to draw on the diagram.

The other symbol is at the top of the diagram and is just a horizontal line marking 6V. This indicates that there will be 6V at that point in the circuit, saving us the trouble of drawing a battery.

Transistor Pinouts Vary

Although this book tries to stick to transistors with compatible pinouts to each other, this is not universal. Not all transistors in a certain package have the same pinout; check on the datasheet before you use a new transistor.

Pinouts for a lot of the devices used in this book are in Appendix A.

Looking at Figure 5-6, you may be wondering why you still need R1, as the gate does not draw any current. The reason it's a good idea to still include the resistor is that when you first raise the gate voltage, there will be a very quick inrush of current for a fraction of a second. The resistor makes sure that this current does not exceed the capabilities of the GPIO pin.

The catch with MOSFETs is that the threshold voltage is sometimes too high to be switched with the 3.3V or 5V of a Raspberry Pi or Arduino. MOSFETs that have a low enough gate threshold to be used directly from a GPIO port are called *logic-level MOSFETs*. This book standardizes on two MOSFETs: for lower power, use the 2N7000; and for higher power applications, use the FQP30N06L. Both have a gate threshold voltage guaranteed to be below 3V, making them suitable for use with both an Arduino and Raspberry Pi.

In general, MOSFETs run a lot cooler when switching loads than bipolar devices. One of the main properties to look for when buying a MOSFET is its *on resistance*. A MOSFET with a very low on resistance will switch high currents without even getting warm. As you might expect, the cost of a MOSFET increases the lower its on resistance.

In this book, you will find MOSFETs being used quite a lot. Mostly I stick to the FQP30N06L (up to 30A), although for this kind of current you will need a big heat sink.

In fact, the emitter, base, and collector pins of the TIP120 Darlington and the source, gate, and drain of an FQP30N06L MOSFET are in the same positions, so you could just unplug a TIP120 from a breadboard and plug in an FQP30N06L using the same configuration and the circuit should still work.

PNP and P-channel transistors

Transistors of all the types just described actually come in two flavors. So far we have only considered one flavor: negative positive negative (NPN) or N-channel in the case of MOS-FETs. These are the most common type and usually that is all you will ever need.

The other flavor of transistors are PNP or P-channel devices. Where N-type devices are used to switch the load to ground, P-type devices switch it to the positive supply. P-channel MOSFETs are used in H-bridge motor drivers in Chapter 8.

Transistor selection guide

Selecting the right transistor for a job can be tricky. Table 5-2 standardizes the choice into just five devices.

When used with a Raspberry Pi, it is assumed that a 1kΩ resistor is between the GPIO pin and the base or the gate of the transistor. For an Arduino, that resistor is assumed to be 150Ω. The current figures are derived by testing on actual devices, and the maximum voltage figures are taken from the devices' datasheets.

Table 5-2 *A useful set of transistors*

Transistor	Type	Package	Max. current (Pi 3.3V)	Max. current (Arduino 5V)	Max. volts
2N3904	Bipolar	TO-92	100mA	200mA	40V
2N7000	MOSFET	TO-92	200mA	200mA	60V
MPSA14	Darlington	TO-92	1A	1A	30V

TIP120	Darlington	TO-220	5A	5A	60V
FQP30N06L	MOSFET	TO-220	30A	30A	60V

When shopping for an FQP30N06L, make sure the MOSFET is the "L" (for logic) version, with "L" on the end of the part name; otherwise, the gate threshold voltage may be too high.

The MPSA14 is actually a pretty universally useful device for currents up to 1A, although at that current there is a voltage drop of nearly 3V and the device gets up to a temperature of 120°C! At 500mA, the voltage drop is a more reasonable 1.8V and the temperature 60°C.

To summarize, if you only need to switch around 100mA then a 2N3904 will work just fine. If you need up to 1A, use an MPSA14. Above that, the FQP30N06L is probably the best choice, unless price is a consideration, because the TIP120 is considerably cheaper.

Diodes

The diode in experiment 1 (see Figure 5-1) is included to protect the Raspberry Pi or Arduino and the transistor.

Motors create voltage spikes and all sorts of electrical noise that can wreak havoc with delicate electronics like the Raspberry Pi or Arduino. A diode is used to make sure that these electrical spikes caused by the motor cannot momentarily reverse the flow of current, something that can easily destroy the transistor. It does this because the diode only allows current to pass in one direction. It is a kind of one-way valve. Current can only flow in the direction indicated by its arrow-like shape.

It is quite common to attach a diode across the terminals of a motor in this way. The current will usually flow through the motor in the opposite direction to the direction allowed by the diode, but should there be a negative spike in voltage, the diode will come into play, conduct and effectively short out the brief flow of current, nullifying it.

LEDs

You will learn much more about LEDs in Chapter 6. The term LED is an acronym for Light Emitting Diode.

As you might guess, an LED works like a plain diode, except that when current is flowing through it, it emits light. The symbol for an LED is the same as for a normal diode, but with arrows on it to indicate the emission of light (Figure 5-7).

Figure 5-7 *The symbol for an LED*

LEDs are available in a wide range of colors and sizes. They can be driven directly from an Arduino or Raspberry Pi GPIO pin, but you must use a resistor to limit the current, as you would when driving a transistor. You will discover how to do this in Chapter 6.

Capacitors

Capacitors can be thought of as places where you can temporarily store electricity—akin to very low capacity batteries that can keep a small reserve of charge. They will be used in this book in various roles, from helping to suppress electrical interference to keeping a small reserve of extra electrical energy so that when there are sudden peaks in demand they can be drawn from a capacitor.

Figure 5-8 shows the symbols for a capacitor, which, if it has a high value of capacitance, will usually be polarized. Low-value capacitors do not have a positive and negative side.

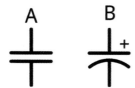

Figure 5-8 *The symbols for a capacitor: unpolarized (A) and polarized (B)*

You may also come across slightly different symbols for capacitors that are recognizable but use a hollow box as the positive end of a polarized capacitor and a solid box as the negative end. This book uses the US convention for capacitor symbols shown in Figure 5-8.

Integrated Circuits

Integrated circuits (ICs), or *chips* are they are often known, are made up of many transistors built onto a single slice of silicon and encapsulated into a single package. The Raspberry Pi and Arduino are each made of many ICs and some other components attached to a printed circuit board (PCB).

There are special-purpose ICs designed for pretty much any kind of electronic use that you could want. Of particular relevance for this book are ICs that help you to control things.

These kind of ICs will often combine high-current transistors and control logic into a single package.

The Ins and Outs of Connections

Having covered the basics of electronics, we should also look at how those electronics interface with a Raspberry Pi or Arduino. You will find much more specific and detailed explanations of this, especially the programming side of things in Chapters 2 and 3.

Digital Outputs

As you saw in Chapter 4, digital outputs are used to turn things on and off. If you are using a Raspberry Pi, a digital output will either be at 0V low or 3.3V high. For an Arduino, the high voltage is 5V rather than 3.3V, but the principle is the same; they cannot be at any voltage in between high and low.

Another difference between the Arduino and the Raspberry Pi is that the Arduino can supply more current (40mA rather than 16mA).

This book is littered with examples that use digital outputs—it is the main mechanism used to control things.

Digital Inputs

Digital inputs are often connected to switches or digital outputs from other devices. A digital input will have a threshold voltage, usually in the middle of the high/low voltage range, so for an Arduino that threshold will be around 2.5V and for a Raspberry Pi it will be around 1.65V. When the program running on the Arduino or Raspberry Pi reads a digital input, if it is above the threshold, it is considered to be high; otherwise, the input is low.

Just like digital outputs, there are no half-measures with digital inputs—they are either high or low.

You can find information on using digital inputs on the Arduino in "Digital Inputs" on page 17 in Chapter 2 and on the Raspberry Pi in "Digital Inputs" on page 39 in Chapter 3.

Analog Inputs

Analog inputs allow you to measure a voltage on an analog pin that lies between the low and high voltages. The Raspberry Pi does not have any analog inputs, but the Arduino has six, labeled A0 to A5.

In an Arduino, the voltage between 0 and 5V is mapped to a number between 0 and 1023. So 0V gives a reading of 0, and 5V a reading of 1023. Something in the middle (2.5V) will give a reading of around 511.

You can find out more about analog inputs on the Arduino in "Analog Inputs" on page 18 in Chapter 2.

Analog Outputs

Although you might imagine analog outputs would allow you to set an output pin to any voltage between low and high, they are a bit more complicated than that. They use a technique called pulse-width modulation (PWM) to control the average power arriving at a normal digital output.

PWM is used to control the speed of motors and the brightness of LEDs. "Controlling Speed (PWM)" on page 96 explains how PWM works, and can be used to control the speed of a motor.

Serial Communication

The interface techniques just described are the basic, low-level techniques. Some devices that you will want to control from an Arduino or Raspberry Pi will use serial interfaces that pass binary data one bit at a time from one device's digital output to another device's digital input.

For example, in Chapter 14, you will find displays that require their data in serial form.

There are various standards of serial interface that all do a very similar job but in slightly different ways. These include what is known as simply *serial* or *TTL serial*, *I2C* (pronounced i squared c), and serial peripheral interface (SPI).

Summary

In this chapter you have learned all about the fundamental electronic concepts of current voltage and resistance, and looked at some electronic components that are used in this book. But that's enough theory for now! In the next chapter, you will learn how to use various types of LED with your Arduino and Raspberry Pi.

LEDs §

Controlling light-emitting diodes (LEDs) is likely to be near the top of any maker's to-do list. There is a bewildering array of different types of LEDs, from simple LEDs, through to high-power, infrared, and ultraviolet devices.

LEDs (Figure 6-1) vary in their current requirements, and some will work just fine from a digital output, while others will need a transistor or other circuitry to control them.

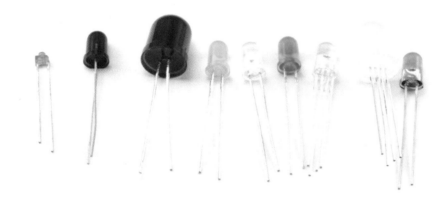

Figure 6-1 *A selection of LEDs*

Regular LEDs

By regular LEDs, I mean those small colorful devices that often come in a 5mm diameter (sometimes 3mm or 10mm) and require a modest current to make them light, so that they can be controlled directly from the output of an Arduino or Raspberry Pi.

Figure 6-2 shows how you typically connect an LED to a digital output of an Arduino or Raspberry Pi.

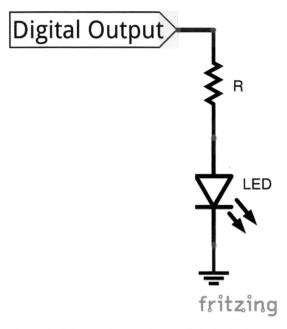

Figure 6-2 *Connecting an LED to a digital output*

The resistor is needed to limit the current flowing through the LED for two reasons: first, so that the LED's maximum current is not exceeded (which will shorten its life); and second, so that the output-current capability of the output pin or the cumulative total of all the output pins has not been exceeded.

Current Limiting

When an LED is connected up as shown in Figure 6-2 there will be a more or less constant voltage across the LED. This is called the LED's *forward voltage* (abbreviated to *Vf*). Vf is different for different colors of LED. Red LEDs usually have the lowest Vf and blue and white LEDs usually have the highest Vf among LEDs producing visible light (see Table 6-1).

There are also infrared (IR) LEDs, which are used in TV remote controls, and ultraviolet (UV) LEDs, which are often used to make white clothes glow a violet color at parties or to check for fake bank notes.

In addition to its forward voltage, the other thing to know about an LED is the forward current that you want flowing through it. Most LEDs will emit some light at a current of 1mA or less, but most will reach optimum brightness at around 20mA. That's quite a big range, and is why if you want to play it safe, a 470Ω resistor with any LED from either a Raspberry Pi or Arduino will work, although the LED will not be as bright as it could be.

The easiest way to calculate the value of series resistor is to use a web service that will do the math for you. One such service (*http://led.linear1.org/1led.wiz*) is shown in Figure 6-3.

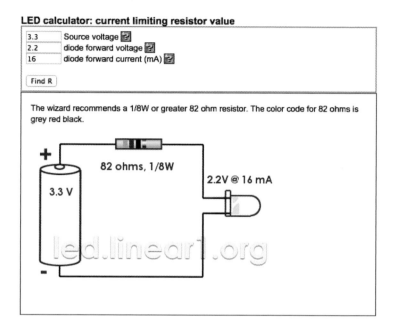

Figure 6-3 *Series resistor online calculator*

In this case, the source voltage is 3.3V because a Raspberry Pi is being used and the Raspberry Pi's maximum current of 16mA for any one pin is used. The calculator tells us that we should use an 82Ω resistor for an LED with a Vf of 2.2V.

If you are interested in making the calculation for yourself, you'll need to first subtract the Vf (2.2V) value for the LED from the logic-level voltage (3.3V). That makes 1.1V. Then use Ohm's law to calculate the resistor value R = V / I = 1.1V / 16mA = 68.75Ω.

You can also use Table 6-1 to help you pick a resistor. This table indicates the approximate range of Vf for different colors of LED as well.

Table 6-1 *Current limiting resistors for LEDs*

	IR	Red	Orange/ Yellow/ Green	Blue/ White	Violet	UV
Vf	1.2-1.6V	1.6-2V	2-2.2V	2.5-3.7V	2.7-4V	3.1-4.4V
Pi 3.3V 3mA	X	680Ω	470Ω	270Ω	220Ω	68Ω
Pi 3.3V 16mA	150Ω	120Ω	82Ω	56Ω	39Ω	15Ω
Arduino 5V 20mA	220Ω	180Ω	150Ω	150Ω	120Ω	100Ω

Note that the resistance values are adjusted to the nearest readily available standard resistor value.

There is an X in the IR column at 3mA because IR LEDs for remote controls generally require at least 10mA to do anything much and most are designed to operate at 100mA or more to obtain a decent range for the remote control.

Project: Traffic Signal

Having such a great selection of LED colors to choose from for this project (shown in Figure 6-4), you will use a red, orange, and green LED to make a traffic signal controlled by Arduino or Raspberry Pi.

Figure 6-4 *An Arduino traffic signal*

The sequence of LEDs displayed is:

1. Red

2. Red and orange together

3. Green

4. Orange

Parts List

Whether you are using a Raspberry Pi or Arduino (or both), you are going to need the following parts to build this project:

Name	Part	Sources
LED1	Red LED	Adafruit: 297
		Sparkfun: COM-09590
LED2	Orange LED	Sparkfun: COM-09594
LED3	Green LED	Adafruit: 298
		Sparkfun: COM-09650
R1-3	150Ω resistors	Mouser: 291-150-RC
	400-point solderless breadboard	Adafruit: 64
	Male-to-male jumper wires	Adafruit: 758
	Female-to-male jumper wires (Pi only)	Adafruit: 826

Over time, you will probably find yourself collecting resistors of many different values. In this case, I have suggested a compromise value of 150Ω resistors for the LEDs for both Arduino and Raspberry Pi and for all three colors of LED. If you wanted to fine-tune this for maximum brightness, you could use Table 6-1 to pick the optimum resistor values.

Design

The three LEDs are each connected to a separate output pin of either the Arduino or Raspberry Pi.

Arduino Connections

Figure 6-5 shows the breadboard layout and connection to the Arduino.

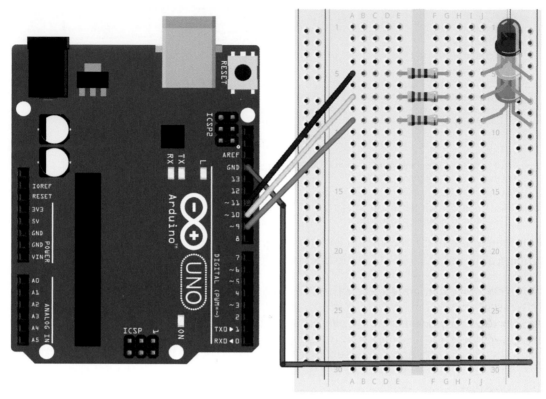

fritzing

Figure 6-5 *Breadboard layout for the Arduino traffic signal*

Remember that the long LED lead is the positive lead and these go to the left of the bread-board and one end of the LED's resistor. The shorter negative lead of the LEDs all go to the negative supply column that runs down the righthand side of the breadboard.

Arduino Software

The Arduino sketch for this project can be found in *arduino/projects/traffic_signals*, which you'll find in the place where you downloaded the book's code (see "The Book Code" on page 14 in Chapter 2). Let's take a look at it:

```
const int redPin = 11;          // ❶
const int orangePin = 10;
const int greenPin = 9;

void setup() {                  // ❷
  pinMode(redPin, OUTPUT);
  pinMode(orangePin, OUTPUT);
  pinMode(greenPin, OUTPUT);
}
```

```
void loop() {                          // ❸
  setLEDs(1, 0, 0);
  delay(3000);
  setLEDs(1, 1, 0);
  delay(500);
  setLEDs(0, 0, 1);
  delay(5000);
  setLEDs(0, 1, 0);
  delay(500);
}

void setLEDs(int red, int orange, int green) {   // ❹
 digitalWrite(redPin, red);
 digitalWrite(orangePin, orange);
 digitalWrite(greenPin, green);
 }
```

The sketch has much in common with the original blink sketch, except that instead of controlling a single LED, three LEDs are being controlled.

❶   Constants are defined for each of the Arduino pins connected to an LED.

❷   In the setup() function, the pins are set to be outputs.

❸   The loop() function calls a setLEDs() function to set the three LEDs on or off (1 or 0). The delay between each call to setLEDs() determines how long the lights stay in that phase.

❹   The setLEDs() function really just makes the loop function shorter and easier to read, allowing all three digitalWrites() to be condensed into one line.

Raspberry Pi Connections

For the Raspberry Pi version of this project, the male-to-male jumpers are swapped for female-to-male jumpers and the same breadboard layout is connected to GPIO pins 18, 23, and 24, as shown in Figure 6-6.

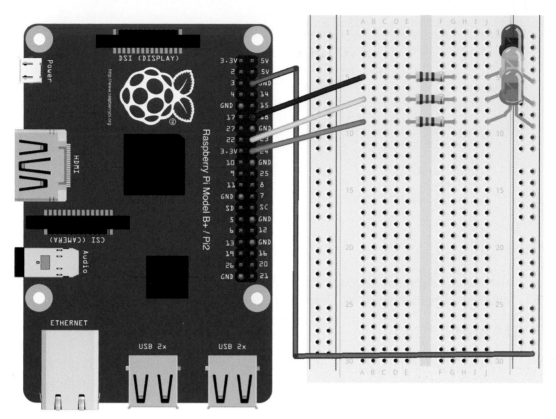

Figure 6-6 *Breadboard layout for the Raspberry Pi traffic signal*

Raspberry Pi Software

The Python program for this project can be found in the *traffic.py* file in the *python/ projects/* directory (for information on installing the Python programs for the book, see "The Book Code" on page 34 in Chapter 3):

```
import RPi.GPIO as GPIO
import time

GPIO.setmode(GPIO.BCM)

red_pin = 18
orange_pin = 23
green_pin = 24

GPIO.setup(red_pin, GPIO.OUT)
GPIO.setup(orange_pin, GPIO.OUT)
GPIO.setup(green_pin, GPIO.OUT)

def set_leds(red, orange, green):    # ❶
```

```
        GPIO.output(red_pin, red)
        GPIO.output(orange_pin, orange)
        GPIO.output(green_pin, green)

    try:
        while True:
            set_leds(1, 0, 0)
            time.sleep(3)
            set_leds(1, 1, 0)
            time.sleep(0.5)
            set_leds(0, 0, 1)
            time.sleep(5)
            set_leds(0, 1, 0)
            time.sleep(0.5)

    finally:
        print("Cleaning up")
        GPIO.cleanup()
```

❶ Just like the Arduino version, a function (set_leds) is used to keep the main loop uncluttered by too many GPIO.output functions.

PWM and LEDs

If you try controlling the brightness of an LED by adjusting the voltage across it, it generally doesn't work so well, because there is a big dead-zone until the voltage gets high enough for the LED to emit any light at all.

PWM analog outputs (see "Pulse-Width Modulation" on page 84) are perfect for controlling the brightness of the LED. LEDs can turn on and off very quickly, certainly less than a millionth of a second, so when using PWM with LEDs, they are actually flashing at the PWM frequency, but the eye sees them as on but with a brightness that varies as the proportion of the time that the LED is actually on changes.

Pulse-Width Modulation

So far, you have controlled things in a very digital way—that is, you have turned things on and then turned them off again. But what if you want to control things in a more analog manner? For example, perhaps you want to control the speed of a motor or the brightness of an LED. To do this, you need to control the amount of power being supplied to the thing you want to control.

The technique used to control power in this way is called pulse-width modulation (PWM) and uses a digital output, producing a series of high and low pulses. By controlling the amount of time that the pulse is high, you can control the overall transfer of power into a motor or LED.

Figure 6-7 shows how PWM works, assuming the output is the 3.3V of a Raspberry Pi GPIO pin. You would also need a transistor connected to the GPIO pin to supply enough current to drive a motor.

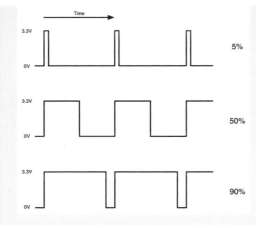

Figure 6-7 *Pulse-width modulation*

motor or LED will receive half power and probably rotate at something close to half speed (or brightness). If the duty cycle is increased to 90%, the motor will be almost at full speed and the LED fully lit.

Turning the motor or LED off is just a matter of setting the duty cycle to 0 and full speed setting it to 100%. Setting the duty cycle to 100% will leave the GPIO pin high all the time.

Both the Arduino and Raspberry Pi have the ability to do PWM on their output pins. On an Arduino Uno, this is restricted to pins D3, D5, D6, D9, D10, and D11. These pins are marked with a ~ sign next to the pin on the Arduino itself.

The frequency of these pulses can be varied on both Arduino and Raspberry Pi. On an Arduino Uno, it is 490Hz (pulses per second) for most pins except pins 5 and 6, which run at 980Hz.

For controlling the brightness of an LED or the speed of a motor, the default Arduino PWM frequency of 490Hz works just fine.

The proportion of the time that the pulses are high is called the duty cycle. If the pulses are only on for 5% of the time (duty cycle of 5%), then not much energy will be arriving at the motor, so it will turn very slowly (or the LED will be dim). Increase the duty cycle to 50% and the

RGB LEDs

Red, green, and blue LEDs (RGB LEDs) are a single LED case that contains three actual LEDs. The three LEDs are red, green, and blue. By using PWM to control the brightness to each of the LED colors, you can make the LED appear any color.

Although an RGB LED contains three normal, two-pin LEDs, this does not mean that the LED body has to have six pins, because one terminal of each LED can be connected together as a common pin (Figure 6-8).

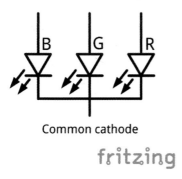

Common cathode

fritzing

Figure 6-8 *The schematic diagram for a common cathode RGB LED*

If the negative connections to each LED are connected together, the resulting lead is called a common cathode and if the positive connections are common, the lead is called a common anode.

The RGB LED package will either be described as clear or diffuse. If it's clear you will be able to see the red, green, and blue LEDs inside the package, and the color will not mix together so well. The diffuse packages mix together the light from the three LEDs much better.

In Chapter 14, you will work with displays made up of RGB LED ICs that contain a chip that both limits the current to the red, green, and blue LEDs, in addition to providing them with a serial data interface that allows an Arduino or Raspberry Pi to control large numbers of them with a single output pin.

Experiment: Mixing Colors

In this experiment, you can use both Arduino and Raspberry Pi to control the color of an RGB LED. In the Raspberry Pi version of the project, a graphical user interface with three sliders is used to control the color. Figure 6-9 shows the Raspberry Pi version of the experiment.

Hardware

Figure 6-10 shows the schematic diagram for the experiment.

Figure 6-9 *Mixing colors with a Raspberry Pi*

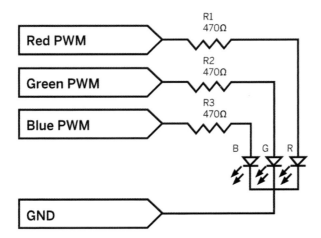

Figure 6-10 *The schematic diagram for the RGB LED experiment*

For optimum brightness and best color balance, you should choose your resistor values carefully. However, it's easier to buy your components if you use the same value resistor for all three channels. In this case, I suggest 470Ω resistors as a "universal" option that will work with Raspberry Pi or Arduino.

The brightness and efficiency of an RGB LED is such that even with only 3mA (6mA on an Arduino), the LED will still look pretty bright.

Parts List

Whether you are using a Raspberry Pi or Arduino (or both), you are going to need the following parts to carry out this experiment:

Name	Part	Sources
LED1	RGB LED diffused common cathode	Sparkfun: COM-11120
R1-3	470Ω resistor	Mouser: 291-470-RC
	400-point solderless breadboard	Adafruit: 64
	Male-to-male jumper wires	Adafruit: 758
	Female-to-male jumper wires (Pi only)	Adafruit: 826

If you are planning to try this experiment with a Raspberry Pi, you will need female-to-male jumper wires to connect the Raspberry Pi GPIO pins to the breadboard.

All the components in the list are included in the MonkMakes Electronic Starter Kit for Raspberry Pi (see Appendix A).

Arduino Connections

The Arduino version of this experiment uses the Serial Monitor to set the proportion of red, green, and blue.

Figure 6-11 shows the breadboard layout and connections to the Arduino.

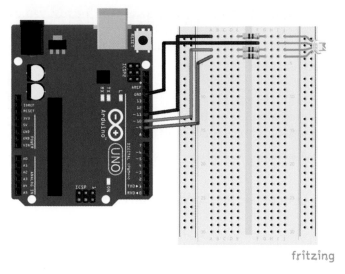

Figure 6-11 *Arduino breadboard layout for RGB color mixing*

The longest lead of the LED is the common cathode lead that is connected to GND in Figure 6-11. The order of the other leads may not be the same as I have shown, so you might have to do some lead swapping if you find your colors mixed up.

Arduino Software

The Arduino sketch for this experiment uses three PWM channels to control the brightness of each of the three colors. The sketch is in the *ex_12_mixing_colors* file, which you'll find in the place where you downloaded the book's code):

```
const int redPin = 11;
const int greenPin = 10;
const int bluePin = 9;

void setup() {
  pinMode(redPin, OUTPUT);
  pinMode(greenPin, OUTPUT);
  pinMode(bluePin, OUTPUT);
  Serial.begin(9600);
  Serial.println("Enter R G B (E.g. 255 100 200)");
}

void loop() {
  if (Serial.available()) {
    int red = Serial.parseInt();    // ❶
    int green = Serial.parseInt();
    int blue = Serial.parseInt();
    analogWrite(redPin, red);       // ❷
    analogWrite(greenPin, green);
    analogWrite(bluePin, blue);
  }
}
```

❶ Each of the three PWM values that should be between 0 and 255 are read into variables.

❷ The PWM output is then set for each channel.

Arduino Experimentation

After uploading the sketch, open the Serial Monitor on the Arduino IDE. Then type three numbers, each between 0 and 255 and separated by a space, and click the Send button.

The LED should then change color to match the red, green, and blue numbers that you just entered.

You can check each channel separately by entering 255 0 0 (red), then 0 255 0 (green), and finally, 0 0 255 (blue).

Raspberry Pi Connections

Rather than just use commands to change color, the Raspberry Pi version of this project creates a small user interface window on the Raspberry Pi that has three sliders, one for each color channel.

As you change the position of the sliders, the LED color will change. This software uses a graphical user interface, so you will need to have a keyboard, mouse, and monitor attached to your Raspberry Pi, as SSH does not have a graphical interface.

The breadboard layout for the Raspberry Pi (Figure 6-12) is just the same as for the Arduino, except that for the Raspberry Pi you need to use female-to-male jumper wires. The GPIO pins 18, 23, and 24 are used as PWM outputs.

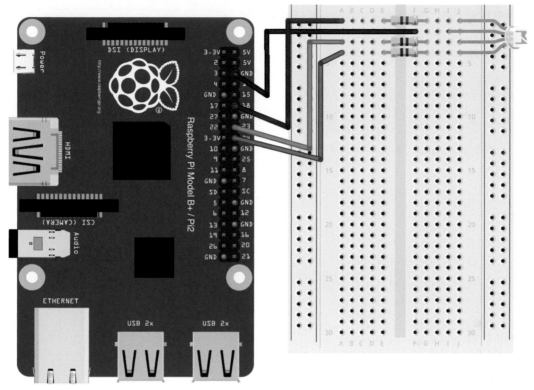

Figure 6-12 *Raspberry Pi breadboard layout for RGB color mixing*

Raspberry Pi Software

The Python program for this experiment uses a framework called Tkinter that allows you to make applications that run in a window and have user interface controls, rather than the simple command line that you have used so far. This makes the program a bit longer than normal. It also makes use of some more advanced programming.

Let's take a look at the program code (which you can find in the file *mixing_colors.py*):

```
from Tkinter import *
import RPi.GPIO as GPIO
import time


GPIO.setmode(GPIO.BCM)  # ❶
GPIO.setup(18, GPIO.OUT)
GPIO.setup(23, GPIO.OUT)
GPIO.setup(24, GPIO.OUT)


pwmRed = GPIO.PWM(18, 500) # ❷
```

```
pwmRed.start(100)

pwmGreen = GPIO.PWM(23, 500)
pwmGreen.start(100)

pwmBlue = GPIO.PWM(24, 500)
pwmBlue.start(100)

class App:

    def __init__(self, master): # ❸
        frame = Frame(master)  # ❹
        frame.pack()

        Label(frame, text='Red').grid(row=0, column=0) # ❺
        Label(frame, text='Green').grid(row=1, column=0)
        Label(frame, text='Blue').grid(row=2, column=0)

        scaleRed = Scale(frame, from_=0, to=100,    # ❻
            orient=HORIZONTAL, command=self.updateRed)
        scaleRed.grid(row=0, column=1)
        scaleGreen = Scale(frame, from_=0, to=100,
            orient=HORIZONTAL, command=self.updateGreen)
        scaleGreen.grid(row=1, column=1)
        scaleBlue = Scale(frame, from_=0, to=100,
            orient=HORIZONTAL, command=self.updateBlue)
        scaleBlue.grid(row=2, column=1)

    def updateRed(self, duty):      # ❼
        # change the led brightness to match the slider
        pwmRed.ChangeDutyCycle(float(duty))

    def updateGreen(self, duty):
        pwmGreen.ChangeDutyCycle(float(duty))

    def updateBlue(self, duty):
        pwmBlue.ChangeDutyCycle(float(duty))


root = Tk()  # ❽
root.wm_title('RGB LED Control')
app = App(root)
root.geometry("200x150+0+0")
try:
    root.mainloop()
finally:
    print("Cleaning up")
    GPIO.cleanup()
```

❶   Configure the Pi to use the Broadcom (BCM) pin names, rather than the pin positions.

❷   Start pulse-width modulation (PWM) on the red, green, and blue channels to control the brightness of the LEDs.

❸   This function gets called when the app is created.

❹   A frame holds the various GUI controls.

❺   Create the labels and position them in a grid layout.

❻   Create the sliders and position them in a grid layout. The `command` attribute specifies a method to call when a slider is moved.

❼   This method and the similar methods for the other colors are called when their slider moves.

❽   Set the GUI running, and give the window a title, size, and position.

Raspberry Pi Experimentation

Run the program as superuser using the following command:

```
$ sudo python mixing_colors.py
```

After a moment or two, the window shown in Figure 6-13 will appear.

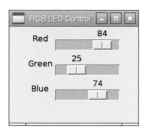

Figure 6-13 *Move the sliders to change the LED color*

When you move the sliders around, the color of the LED will change.

Summary

In this chapter you have learned how to switch an LED on and off and also control its brightness using an Arduino and Raspberry Pi. Now that you can control light with the Arduino and Raspberry Pi, the next chapter will focus on movement. We'll explore DC motors, which are the most common type, and how we can control them.

Motors, Pumps, and Actuators | 7

We started experimenting with DC motors in Chapter 4. A lot of the principles that you will learn when using DC motors are applicable to many things that you might want to control with an Arduino or Raspberry Pi.

Figure 7-1 shows a selection of DC motors. As you can see, these devices come in all sorts of shapes and sizes.

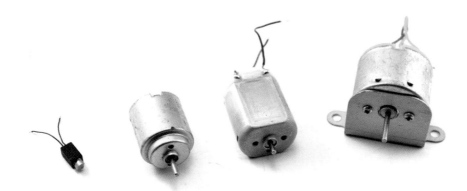

Figure 7-1 *A range of DC motors*

Motors are also the driving force behind lots of other useful output devices like pumps and linear actuators, both of which you will meet later in this chapter.

As you discovered in "Experiment: Controlling a Motor" on page 53, DC motors need too much current to drive them directly from the output pin of a Raspberry Pi or Arduino, but can be switched on and off using a transistor.

In this chapter, you will start by controlling the speed of a DC motor.

How DC Motors Work

DC motors generally have three major components, as shown in Figure 7-2. They have stationary magnets (stator) around the outside of the motor, a rotor (the bit that moves) and a commutator.

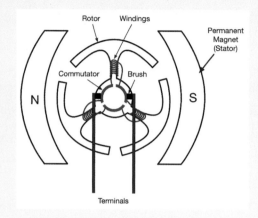

Figure 7-2 *The insides of a DC motor*

Coils of wire are wound around the rotor. In the case of Figure 7-2, there are three coils around three shaped parts of the rotor. These coils are connected to the commutator. The commutator's job is to energize the coils with the right polarity, in turn, as the rotor rotates, so that the next coil is always being pushed and pulled against the permanent magnets in the stator, so that the overall effect is that the rotor rotates.

The commutator is made up of a ring split into segments (in this case, three) and brushes are used to connect the terminals to the separate segments of the commutator as it rotates with the rotor.

This design with three sets of windings is quite typical for a small DC motor like the ones shown in Figure 7-1.

A useful feature of DC motors is that when you reverse the polarity of the voltage across its terminals, it will spin in the opposite direction.

Controlling Speed (PWM)

In "Experiment: Mixing Colors" on page 86, you used pulse-width modulation (see "Pulse-Width Modulation" on page 84 in Chapter 6) to control the brightness of an LED. You can use just the same approach to control the speed of a motor.

Experiment: Controlling the Speed of a DC Motor

This experiment uses exactly the same hardware as "Experiment: Controlling a Motor" on page 53, but instead of just turning the motor on and off, it controls the motor's speed.

Hardware

If you have not already done so, build the hardware for "Experiment: Controlling a Motor" on page 53. As a reminder, the breadboard layout for the hardware is shown in Figure 7-3.

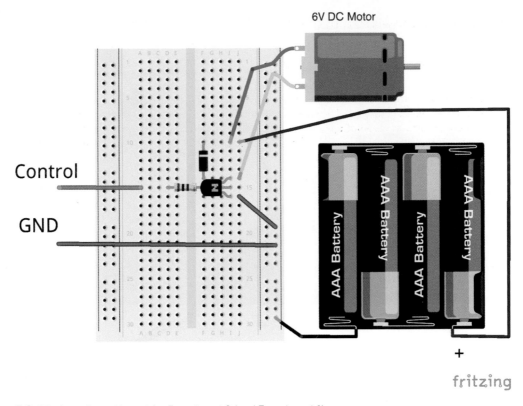

Figure 7-3 *The breadboard layout for Experiment 2 (and Experiment 1)*

Arduino Connections

Connect up the Arduino as shown in Figure 7-4. Connect GND on the breadboard to GND and the Control lead on the breadboard to pin D9 of the Arduino Uno.

Figure 7-4 *Connecting the Arduino to the breadboard*

Arduino Software

The Arduino sketch for this project is in the */experiments/pwm_motor_control* directory, which you can find in the place where you downloaded the book's code (see "The Book Code" on page 14 in Chapter 2).

The program uses the Arduino IDE's Serial Monitor to allow you to enter a duty cycle and then sets the motor speed accordingly:

```
const int controlPin = 9;

void setup() {                        // ❶
  pinMode(controlPin, OUTPUT);
  Serial.begin(9600);
  Serial.println("Enter Duty Cycle (0 to 100)");
}

void loop() {                         // ❷
  if (Serial.available()) {           // ❸
    int duty = Serial.parseInt();
    if (duty < 0 || duty > 100) {     // ❹
      Serial.println("0 to 100");
    }
    else {
      int pwm = duty * 255 / 100;
      analogWrite(controlPin, pwm);
      Serial.print("duty set to ");
      Serial.println(duty);
    }
```

```
    }
  }
```

❶ The setup function defines the controlPin as being an output and also starts serial communication using Serial.begin so that you can send values of duty cycle to the Arduino from your computer.

❷ The loop() function checks for any waiting serial communication that has arrived over USB using Serial.available.

❸ If there is a message, in the form of a number, the number is read from the stream of characters and converted into an int using parseInt.

❹ The value is checked to see if it's in the range 0 to 100. If it's not in this range, a reminder is sent back to your Serial Monitor over USB.

Text and Numbers

We will use this technique of reading numbers from the Arduino Serial Monitor quite a lot in this book, so its worth explaining what parseInt actually does and what's going on when we send a message to the Arduino over USB.

In "Serial Communication" on page 74, I mentioned the fact that one way of interfacing with Arduino (and for that matter, Raspberry Pi) is to communicate with a serial interface. The Arduino Uno has a serial interface attached to digital pins D0 and D1. These pins should not be used as general digital input and output, because they are the serial interface from the Arduino to your computer via a USB interface chip that converts between USB serial and the kind of direct serial that the Arduino understands.

When you type a message in the Serial Monitor and send to the Arduino, the text of the message is converted into a stream of bits (high and low signals) that are reconstructed into groups of eight bits (bytes) as they are received. Each of these bytes is a numeric code for a letter of the Roman alphabet. All the letters and digits have a unique code defined by a standard called ASCII (American Standard Code for Information Interchange).

When there is a message received as serial by the Arduino program that is running, it can either read a single byte (letter) at a time, or it can use the parseInt function that keeps reading characters and as long as the character is a digit, it constructs a number from it. For example, the number 154 would be sent as three characters (that is, 1, 5, and 4). After the character, if there is a newline character, a space, or a character that is not a digit, parseInt knows that all the letters for the number have been received and returns a value of 154 as an int. In fact, even if there is just a short delay after the last letter is sent, that is sufficient for parseInt to decide that that's the end of the number.

On the other hand, if the number is between 0 and 100, then that value is converted into a number between 0 and 255 and then the analogWrite() function used to set the PWM value. This is necessary, because the Arduino analogWrite function accepts a value of duty cycle between 0 and 255, where 0 is 0% and 255 is 100% duty cycle.

Arduino Experimentation

On the computer from which you programmed your Arduino, open the Serial Monitor from the Arduino IDE. To open the Serial Monitor, click the magnifying glass icon in the top right of the Arduino IDE. This is circled in Figure 7-5.

This will then prompt you to enter a value of duty cycle between 0 and 100. Experiment by entering different values to see how this affects the speed of the motor.

Notice that when you get down to say 10 or 20, the motor may not actually turn, but instead make a rather strained kind of whine as there is not quite enough power arriving at the motor to overcome friction and get it moving.

If you intend to use the motor for something practical, this sketch is a great way of working out the useful minimum value of duty cycle for the motor.

Figure 7-5 *Using the Serial Monitor to control the motor speed*

When you want to stop the motor, just set the duty to 0.

Raspberry Pi Connections

Connect the breadboard to your Raspberry Pi, as shown in Figure 7-6. Using female-to-male jumper wires, connect GND of the breadboard to one of the GND pins of the GPIO connector and the control wire of the breadboard to pin 18 of the Raspberry Pi.

Figure 7-6 *Connecting the Motor Control breadboard to a Raspberry Pi*

Raspberry Pi Software

The Raspberry Pi software follows a very similar pattern to the Arduino code. In this case, the program will prompt you for a value of duty cycle and control pin 18 accordingly.

You will find the code for the following program in the *pwm_motor_control.py* file, which is in the *python/experiments* directory (you'll find this in the place where you downloaded the book's code—see "The Book Code" on page 34 in Chapter 3):

```python
import RPi.GPIO as GPIO
import time

GPIO.setmode(GPIO.BCM)

control_pin = 18      # ❶

GPIO.setup(control_pin, GPIO.OUT)
motor_pwm = GPIO.PWM(control_pin, 500)   # ❷
motor_pwm.start(0)                       # ❸

try:
    while True:                          # ❹
        duty = input('Enter Duty Cycle (0 to 100): ')
        if duty < 0 or duty > 100:
            print('0 to 100')
        else:
            motor_pwm.ChangeDutyCycle(duty)

finally:
```

```
print("Cleaning up")
GPIO.cleanup()
```

❶ The first part of the program is the same as the code for "Experiment: Controlling a Motor" on page 53, but after defining pin 18 to be an output.

❷ This line sets up that pin to be a PWM output. With the RPi.GPIO library, you can set any of the GPIO pins to be PWM outputs. The parameter 500 sets the PWM frequency to 500Hz (pulses per second).

❸ The PWM output does not actually start until start is called. Its parameter is the initial duty cycle and because we want the motor to be off initially, this is set to 0.

❹ Inside the main loop, the value of duty is requested from you using input and then it's range checked. If its between 0 and 100, it is used to set the duty cycle using the function ChangeDutyCycle.

Raspberry Pi Experimentation

Run the program as superuser using sudo and then try entering different values of duty cycle. The motor speed should change just as it did with the Arduino:

```
pi@raspberrypi ~/make_action/python $ sudo python pwm_motor_control.py
Enter Duty Cycle (0 to 100): 50
Enter Duty Cycle (0 to 100): 10
Enter Duty Cycle (0 to 100): 100
Enter Duty Cycle (0 to 100): 0
Enter Duty Cycle (0 to 100):
```

Controlling DC Motors with a Relay

If you only need your Arduino or Raspberry Pi to occasionally turn a motor on and off, then one approach is to use a relay. Although this is often considered a rather old-fashioned way of doing things, it has a number of advantages:

- It's easy to do, requiring few components.
- Extremely good isolation between the electrically noisy, high-current motor and the delicate Pi or Arduino.
- High-current handling (with the right relay).
- You can use ready-made relay modules directly with a Raspberry Pi or Arduino.

The downsides to using a relay or a relay module include the following:

- They are relatively large components.
- They can only switch the motor on and off—they cannot control the speed.

- They are electromechanical devices and typically are expected to survive perhaps 10,000,000 switching operations before something in them breaks.

Electromechanical Relays

Figure 7-7 shows perhaps the most commonly used type of relay, often called a *sugar cube relay*, because they are the shape of a sugar cube (sort of) although not the same color, generally being black.

The general principle of an electromechanical relay like this is that when a current (around 50mA) passes through its coil, it acts as an electromagnet and pulls two switch contacts together so that a connection is made. These contacts can be high current and high voltage, able to switch tens of amps.

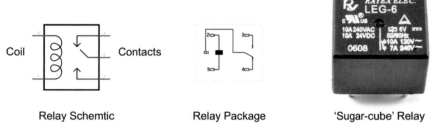

| Relay Schemtic | Relay Package | 'Sugar-cube' Relay |

Figure 7-7 *Relays*

Although you are using a relay to control a motor in this chapter, the relay does just act as a switch, so you can use it to control pretty much anything.

Relays like this are called Single Pole Change Over (SPCO), because rather than have just two terminals that are either connected or not, they actually have three. They have what is called a common terminal, usually labeled COM, and two other terminals: normally open (NO) and normally closed (NC). In this context, normally means without power applied to the relay coil. So, the NO and COM connections will be "open" (not connected) until the coil is energized. The NC contact will behave in the opposite manner, so the NC and COM terminals will "normally" be connected, but will disconnect when the coil is energized.

Generally speaking, just the NO and COM terminals of the relay are used for switching.

Switching a Relay with Arduino or Raspberry Pi

When using a relay with a Raspberry Pi or Arduino, use one that has a 5V coil voltage. Relays coils take too much current (about 50mA) to drive directly from either a Raspberry Pi or an Arduino, and so, in both cases, we will use a small transistor to switch the relay coil to 5V.

Figure 7-8 shows the schematic diagram for switching a relay.

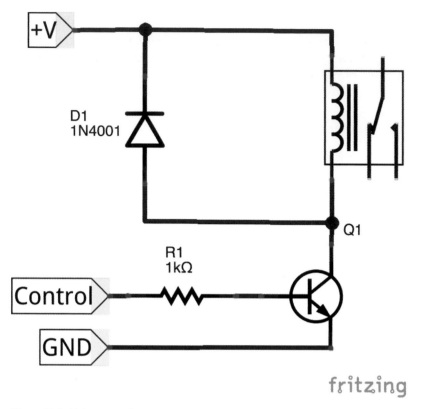

Figure 7-8 *Using a small transistor to switch a relay*

The coil of a relay designed to work at 5V will draw about 50mA of current. This is slightly too much for an Arduino and much too much for a Raspberry Pi GPIO pin. So, just as in "Experiment: Controlling a Motor" on page 53, you will use a transistor to control the motor (in this case, the relay coil will replace the motor as the "load."

This arrangement only makes sense if the motor (or whatever other load you want to control) is of sufficiently high current consumption that it could not be controlled directly using the transistor.

Like a motor, the coil of a relay is capable of producing voltage spikes when it is switched on and off, and so this schematic also needs a diode.

It is very apparent from Figure 7-8 that the switch part of the relay is completely isolated electrically from the coil part. This means that there is less chance of any electrical noise, voltage spikes, or general bad electrical behavior finding its way back to your Arduino and Raspberry Pi.

Because the relay only requires about 50mA when it's powered up, a lowly 2N3904 transistor costing a few cents will work just fine.

Relay Modules

If you have a few things that you want to control and they are compatible with the limitations of relays that I described earlier (on/off control only), a great shortcut is to buy a relay module such as the one shown in Figure 7-9.

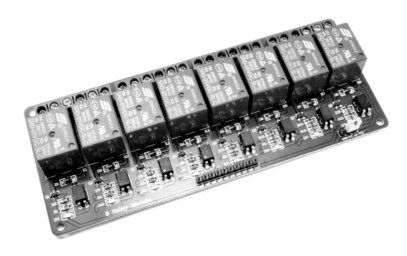

Figure 7-9 *An 8-channel relay module*

These modules are available on eBay and Amazon at very low cost and include both the relay and transistors to drive them, as well as little LEDs that actually tell you when a particular relay has been activated. This means that you can connect them directly to a Raspberry Pi or Arduino.

They are available for controlling a single relay through to controlling even more than the eight that this module has.

The modules will normally have the following pins:

GND (Ground)

VCC or 5V Connect this to 5V on the Raspberry Pi or Arduino. This supplies power to the relay coils when activated.

Data pins Each data pin controls one of the relays. Sometimes these pins will be "active high," meaning you have to set the GPIO pin to which they are connected to high to turn them on and sometimes they are "active low" and the relay coil will be activated when the pin is low.

The module will also have a row of screw terminals that are connected straight to the relay contacts.

Experiment: Controlling a DC Motor with a Relay Module

In this experiment, you will just turn a motor on and off using a relay.

This project uses a ready-made relay module. This module only needs one relay. The relay module I used actually has eight relays, so if you prefer you could use a module with fewer relays on it.

Parts List

Whether you are using a Raspberry Pi or Arduino (or both), you are going to need the following parts to carry out this experiment:

Part	Sources
Small 6V DC motor	Adafruit: 711
Relay module	eBay
6V (4 x AA) battery box	Adafruit: 830
Jumper wires to suit the relay module	See Appendix A

Some relay modules have sockets and some plugs to connect to your Arduino or Raspberry Pi, so select jumper wires to suit. That is, to connect a relay module with male pins to an Arduino, you will need female-to male-jumper wires (Adafruit: 826). To connect it to a Raspberry Pi, you will need female-to-female jumper wires (Adafruit: 266).

Wiring

The wiring diagram for this experiment is shown in Figure 7-10.

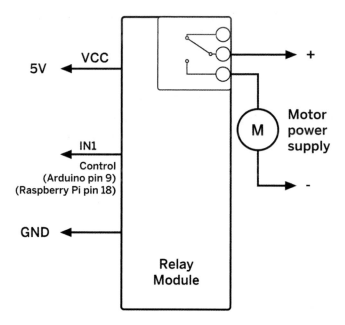

Figure 7-10 *Wiring diagram for controlling a DC motor with a relay module*

The relay contacts act like a switch that can be in one of two positions. There are *common*, *normally open*, and *normally closed* terminals. When the relay coil is not energized, the common terminal is connected to the normally closed terminal. When power is applied to the coil, the switch flips over and the common terminal is now connected to the normally open contact.

The software for both Arduino and Raspberry Pi is almost identical to that of "Experiment: Controlling a Motor" on page 53. The only difference will be if your relay module has active low logic like the one I used.

Arduino Software

The Arduino sketch is in the *arduino/experiments/relay_motor_control* directory (you'll find this in the place where you downloaded the book's code—see "The Book Code" on page 14 in Chapter 2).

The program will turn the relay (and hence the motor) on for 5 seconds, then off for 2 seconds and then repeat. The full code is shown here:

```
const int controlPin = 9;

void setup() {
  pinMode(controlPin, OUTPUT);
}

void loop() {
  digitalWrite(controlPin, LOW);    // ❸
```

```
    delay(5000);
    digitalWrite(controlPin, HIGH);
    delay(2000);
}
```

❶ This code is just the same as for "Experiment: Controlling an LED" on page 46, except that LOW and HIGH are swapped over on the digitalWrite functions. If you find that when you run the program the motor turns on for 2 seconds and off for 5, rather than the other way around, this means that the logic for your relay module is active high and you should swap LOW and HIGH back to how they were in "Experiment: Controlling an LED" on page 46.

Raspberry Pi Software

You will find code for the following program in the *relay_motor_control.py file*, which is in the *python/experiments* directory (you'll find this in the place where you downloaded the book's code):

```
import RPi.GPIO as GPIO
import time

GPIO.setmode(GPIO.BCM)

control_pin = 18

GPIO.setup(control_pin, GPIO.OUT)

try:
    while True:
        GPIO.output(control_pin, False)
        time.sleep(5)
        GPIO.output(control_pin, True)
        time.sleep(2)

finally:
    print("Cleaning up")
    GPIO.cleanup()
```

Choosing a Motor

DC motors come in all sorts of shapes and sizes, and it is important to get one that is powerful enough for your project. The two main things that you need to know about a motor are how much turning force the motor can produce (torque) and how fast the motor spins.

If it spins faster than you need but does not have enough torque, you can trade one off against the other using a gearbox.

Torque

Put crudely, torque is the turning power of a motor. The higher a motor's torque, the stronger its turning force.

Scientifically speaking, torque is defined as a force multiplied by a distance, where the force is measured in newtons (N) and the distance in meters (m). In more practical terms, the force component of torque is often expressed as the force needed to lift a certain weight in kilograms or ounces with a distance measured in centimeters or inches.

Distance comes into the equation because the further away from the motor's rotor, the less force the motor will be able to exert. For example, if a motor is specified as having a torque of 15oz.in (ounces x inches), then 1in from the rotor's center it will be able to hold a weight of 15oz—that is, the motor will not be able to lift the weight higher, but at the same time, the weight won't fall. However, at 10in from the rotor, it will only be able to hold a weight of 1.5oz (10in x 1.5oz = 15oz.in).

Figure 7-11 shows the relationship between weight and distance from the motor's axis.

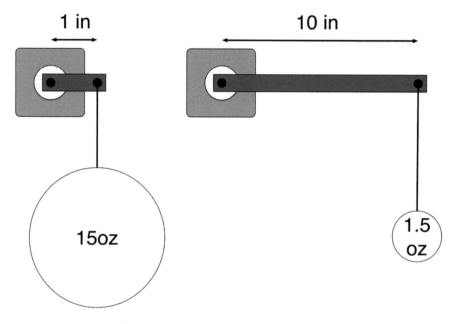

Figure 7-11 *Torque in a nutshell*

RPM

DC motors spin pretty fast, which is why they are often used with gears or as gearmotors (see the following two sections). A typical low-voltage DC motor might spin at 10,000 rpm (revolutions per minute). This means that the motor shaft turns some 166 times per second.

Although you can control the speed of the motor using PWM, this also reduces the energy of the motor, so the torque that the motor generates will remain low.

Gears

Gears allow you to produce a slower rotation with the side effect of increasing torque. If you use a 5:1 gearbox (Figure 7-12) with 50 teeth on one cog and 10 on the other, then for each five turns of the motor, there will be just one turn from the output of the gearbox, but the torque available at the output of the gearbox will be 10 times that available directly from the motor.

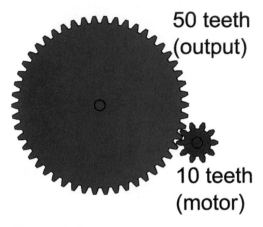

50 teeth (output)

10 teeth (motor)

Figure 7-12 *Gears*

Gearmotors

Because motors are so often used with gears, for many applications, it is better to buy a gearmotor, which combines a motor with a gearbox into a single package.

For examples of a wide range of gearmotors, take a look at Polulu (*https://www.pololu.com/*).

You can buy some extremely cheap gearmotors that use plastic cogs. These work but will not last as long or provide as much torque as a gearmotor with metal gears.

Pumps

Pumps are usually a DC motor, or sometimes a brushless DC motor (these will be discussed in "Brushless DC Motors" on page 201) that drives a mechanism that moves liquid from one place to another.

Two common types of pump that are often used by hobbyists are the peristaltic pump and the velocity pump. These are shown side by side in Figure 7-13. The peristaltic pump is on the left.

Figure 7-13 *A peristaltic pump (left) and velocity pump (right)*

Both of these pumps are powered by a DC motor, but they have quite different properties. If you want slow and measured, use a peristaltic pump; if you want fast and furious, use a velocity pump.

Peristaltic Pumps

Peristaltic pumps are designed to move liquid in a very measured way. In fact, they are often used in medical and research applications to move a fairly precise quantity of liquid. For more accurate control of flow, peristaltic motors are sometime driven by stepper motors (see Chapter 10).

Figure 7-14 shows how a peristaltic pump works.

The pump uses a gear motor driving rollers that squeeze flexible tubing, pushing the fluid through the tubing. It's not surprising that given the constant squeezing, the tube will eventually give out and need replacing. The pumps are usually designed to allow the tubing to be swapped out.

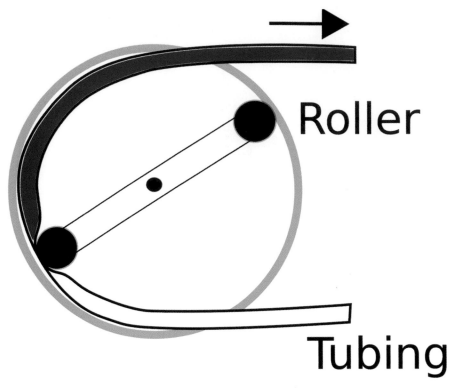

Figure 7-14 *How a peristaltic pump works*

If you drive a peristaltic pump's gearmotor with PWM and an H-bridge (we'll discuss H-bridges in Chapter 8), you can control both the flow rate and direction of flow of the liquid.

A peristaltic pump will self-prime—that is, if the pump is a little higher up than the source of water, it will create enough suction to pull the water into the pump and start pumping it.

Volumetric Flow Rate

Volumetric flow rate is the quantity of liquid that a pump can move in a unit of time. Various units are used in this for both the volume of liquid and the unit of time. A small peristaltic might have a volumetric flow rate of 50 mL/min (millilitres / minute). A velocity pump designed for garden water features might have a volumetric flow rate of 5L/min (5 litres/minute).

Velocity Pumps

If you are more concerned with shifting a lot of liquid quickly, then you need a velocity pump. There are various designs of velocity pump, but the most common is probably the radial pump (Figure 7-15).

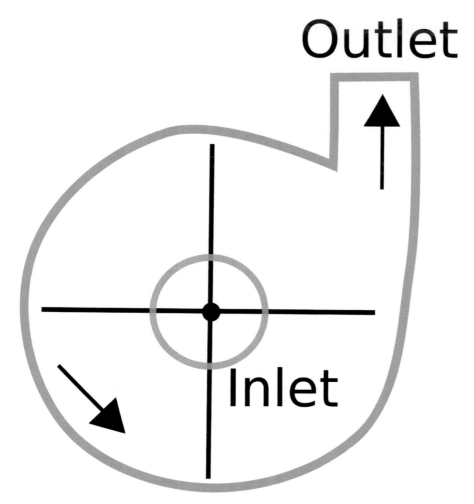

Figure 7-15 *How a radial velocity pump works*

Liquid enters the pump from the front of Figure 7-15 on the axis of the motor which drives impellers. These impellers impart centrifugal force to the liquid that flies out to the edge of the pump housing and out of the outlet.

Velocity pumps are not self-priming, and need to already have water at the inlet before they will pump. Unlike peristaltic pumps, they also allow water to flow through them, even when not pumping. Some of these pumps are intended for garden pond or aquarium use and can be completely submerged in water.

Unlike the peristaltic pump, this type of pump cannot be reversed. Some pumps of this design use a brushless DC motor with its own control electronics all in one enclosure to provide maximum pumping power for a small size.

Project: Arduino House Plant Waterer

This simple Arduino project (Figure 7-16) uses a peristaltic pump to deliver a measured amount of water to your house plants every day (which is very useful while you are away on vacation).

Figure 7-16 *A house plant waterer*

The project does not use a timer to decide when to water the plant; instead it measures the light intensity so that the plant is watered whenever it goes dark.

Design

Figure 7-17 shows the schematic diagram for the project.

An MPSA14 transistor is used by the Arduino to turn the pump's motor on and off. The diode D1 offers protection against negative voltage spikes.

On the left of the schematic, you can see a photoresistor and fixed-value resistor form a voltage divider to measure the light intensity on analog pin A0 of the Arduino.

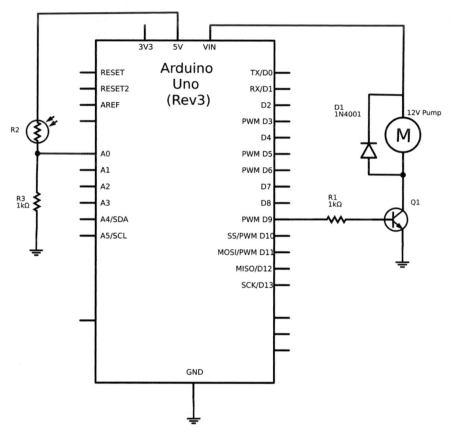

Figure 7-17 *Schematic diagram for the house plant waterer*

The more light there is falling on the photoresistor, the lower its resistance, pulling the voltage on A0 up toward 5V.

This is a pretty easy project to build. Your pump is unlikely to have leads attached, so the only soldering that you may need to do for this project is to attach some leads to the pump motor.

Parts List

You will need the following parts to build this project:

Name	Part	Sources
	Arduino Uno	
Q1	MPSA14 Darlington transistor	Mouser: 833-MPSA14-AP
R1, R3	1kΩ resistor	Mouser: 291-1k-RC
R2	Photoresistor (1kΩ)	Adafruit: 161

		Sparkfun: SEN-09088
D1	1N4001 diode	Adafruit: 755
		Sparkfun: COM-08589
		Mouser: 512-1N4001
	12V peristaltic pump	eBay
	400-point solderless breadboard	Adafruit: 64
	Male-to-male jumper wires	Adafruit: 758
	Tubing to fit the pump 3 ft	Hardware store
	12V 1A power supply	Adafruit: 798
	Large water container	
	Hookup wires to solder to the pump motor	Adafruit: 1311

The type of peristaltic pump used in this project is intended for use in an aquarium, and is available at very low cost.

The tubing that I used in the project came as part of a "watering" kit from a hardware store. It came complete with small plastic joining pieces intended to join one length of the piping they supplied to another. These were perfect for attaching the tube from the pump to the extra lengths of tube. The tubes need to be airtight, or the pump will not work.

Construction

Constructing this project requires a bit of simple work with the breadboard and some possible handiwork to adapt the water container.

Step 1: Solder leads to the motor

Solder leads onto the pump's terminals, unless, of course, it already has leads attached. The leads will need to be long enough to reach from the pump to wherever the breadboard and Arduino will be sited. A length of 18 inches will be plenty long enough.

Step 2: Construct the breadboard

Use the breadboard layout of Figure 7-18 as a reference while you are fitting the components onto the breadboard.

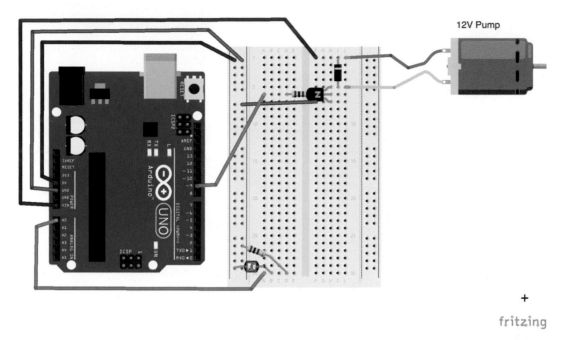

Figure 7-18 *The breadboard layout for the house plant watering project*

Make sure that the transistor and diode are the right way around.

Step 3: Fix tubing onto the pump

You need two lengths of tubing. One about the length of the water container that will reach from the pump inlet, positioned at the top of the container down into the water. The second length needs to reach from the outlet of the pump to the plant that you want to water. Figure 7-19 show a close-up of the pump and the tubes.

The inlet and outlet of the pump are not usually marked, but peristaltic pumps are reversable, so if you find that it's sucking when it should be blowing, it may be easier to swap over the leads to the motor rather than swap over the pipes.

Figure 7-19 *Connecting tubes to the pump*

Step 4: Final assembly

I found it convenient to fit the pump onto the top of the water container with the head of the pump pushed into the top of the container and the motor end sticking up. The inlet tube goes straight down into the container and the outlet tube comes out to the side and off to the plant. To get the pump to fit I had to cut a little of the neck off the milk container that I used.

Figure 7-16 shows the arrangement I used for the project, but if you prefer, you could site the pump low down near the breadboard, but you will need a longer tube.

Software

The sketch for this project is *arduino/projects/waterer/pr_01_waterer.ino*, which you'll find in the place where you downloaded the book's code (you will also find a second sketch in the *projects/* folder called *waterer_test*, which is used to calibrate the light sensor):

```
const int motorPin = 9;       ❶
const int lightPin = A0;

const long onTime = 60 * 1000; // 60 seconds  ❷
const int dayThreshold = 200;          ❸
const int nightThreshold = 70;

boolean isDay = true;                   ❹

void setup() {
  pinMode(motorPin, OUTPUT);
}

void loop() {          ❺
```

```
    int lightReading = analogRead(lightPin);
    if (isDay && lightReading < nightThreshold) { // it went dark ❻
      pump();
      isDay = false;
    }
    if (!isDay && lightReading > dayThreshold) {     ❼
      isDay = true;
    }
  }

void pump() {     ❽
  digitalWrite(motorPin, HIGH);
  delay(onTime);
  digitalWrite(motorPin, LOW);
}
```

❶ The sketch starts by defining constants for the two Arduino pins that are used: the motor control pin and the analog input (lightPin) that uses the photoresistor to measure the light intensity.

❷ The constant onTime specifies how long the pump should stay on each night. While you are testing the project, you might want to set this to a short period, perhaps 10 seconds to prevent too much waiting around.

The most interesting part of this sketch is the part that detects nightfall. Because the Arduino does not have a built-in clock, it does not know the time unless you add real time clock (RTC) hardware to it. For the purposes of this project, we want the plant to be watered once a day, so nightfall will be an appropriate trigger for the watering to start. Having watered the plant, you don't then want watering to be triggered until the next night, so an intervening period of brightness (otherwise known as day) is needed.

❸ To help distinguish between night and day, two constants are defined: dayThreshold and nightThreshold. You will probably need to change these values to suit the location of your plant and the sensitivity of your photoresistor. The basic idea is that if the current light reading is greater than dayThreshold it's daytime and if it's less than nightThreshold, then it's nighttime. You might be wondering why there are two constants rather than just one. The answer is that at dusk, when it just starts to go dark, the light level might increase above and below the threshold for a while, causing multiple triggerings.

❹ The Boolean variable isDay holds the current state of daytime-ness. If isDay is true, then as far as the plant waterer is concerned, it's daytime.

❺ The logic to decide if a watering is due is in the loop function. This first takes a light reading.

❻   If it is currently daytime but the light reading is below `nightThreshold` then it's just gone dark and so the function pump is called to do some watering. The variable `isDay` is then set to `false` to indicate that it's nighttime and prevent further watering for now.

❼   The second `if` statement in `loop` checks that it's nighttime (`!isDay`) and that the light level is now above `dayThreshold`. If both these conditions are true, then `isDay` is set to be `true`.

❽   Finally, the `pump` function has the job of turning on the pump, waiting for the period specified in `onTime` and then turning the pump off again.

Using the Project

Before running the project proper, the test program *waterer_test.ino* should be uploaded to the Arduino, so that you can find suitable values for `dayThreshold` and `nightThreshold`. So upload this sketch to the Arduino and open the Serial Monitor.

A series of readings should appear, one every half second in the Serial Monitor. These correspond to the current light reading. Make a note of the reading during daylight, when it's fairly overcast. Use about half this value for `dayThreshold`, to allow for really dull days.

Next, wait until it goes as dark as it's going to get at the plant's location and take another reading. You may prefer to just guess, or put your finger over the light sensor. Use twice this reading for `nightThreshold`. Remember that you do need `nightThreshold` to be significantly less than `dayThreshold`. So you may need to make compromises in your estimates for these two values.

Now you can alter `dayThreshold` and `nightThreshold` in the real sketch (*pr_01_waterer.ino*) and then upload it to the Arduino.

You can simulate darkness falling by putting your finger over the photoresistor. This should start the pump running for the period specified.

The pump I used pumped about 90mL/minute. To decide on a watering time, you can use a measuring jug and stopwatch to find your pump's volumetric flow rate and adjust `onTime` to deliver the amount of water your plant needs.

Linear Actuators

Linear actuators convert the rotation of a DC motor into a linear movement. They are often used to open and close doors or windows.

They use a threaded drive shaft with what's effectively a nut on it that is constrained from turning, but is free to move up and down the threaded shaft, pushing the end of the actuator in and out. Figure 7-20 illustrates how this works, and Figure 7-21 shows a typical linear actuator.

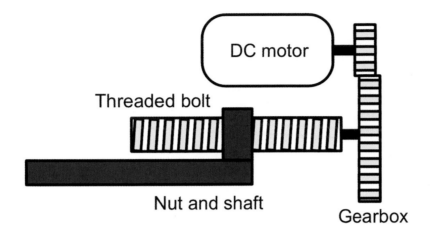

Figure 7-20 *How a linear actuator works*

Figure 7-21 *A linear actuator*

Linear actuators move the shaft quite slowly as the long threaded shaft and nut combination effectively acts as a reduction gearbox, and in any case there is usually a normal gearbox on the end of the motor as well. This low speed and the usually quite powerful DC motors attached mean that they can provide quite a strong pushing or pulling force. The one shown in Figure 7-21 has a pushing or pulling force of 1500N (newtons). This is enough to lift 330lbs or 150kg. At full load, the motor on a linear actuator like this can easily draw 5A at 12V.

You would generally drive the motor for a linear actuator with an H-bridge (at least 5A maximum current) to allow it to move in both directions. To prevent the linear actuator damaging itself when it gets to either end of its travel, the devices usually incorporate limit switches that cut off the power automatically when the shaft reaches an end stop. This simplifies driving the motor, because you can just instruct the H-bridge to power the motor in one direction or the other for a fixed time period that is long enough for the actuator to extend or retract fully.

In "Project: Arduino Beverage Can Crusher" on page 149, a linear actuator like the one in Figure 7-21 is used to make a beverage can crusher.

Solenoids

Solenoids are used for door latches and valves. Like linear actuators, they have a linear motion, but they are much simpler devices. They are effectively electromagnets that make an armature move in and out. The range of travel is very short, usually a fraction of an inch. Figure 7-22 illustrates how a solenoid works, and Figure 7-23 shows a 12V valve that uses a solenoid.

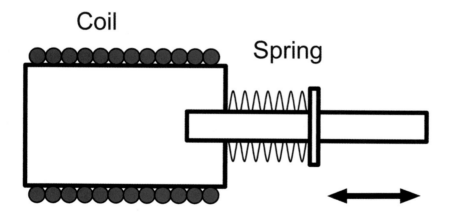

Figure 7-22 *How a solenoid works*

Figure 7-23 *A 12V water control valve*

When the coil is energized, it pulls the plunger into the coil against the force of the spring. When the power is removed from the coil, the plunger is free to move back to its initial position.

The water valve shown in Figure 7-23 is designed to switch a pressurized household water supply. When no power is supplied to the valve, no water flows, but when the coil is energized, a plunger retracts, allowing water to flow through the valve for as long as power is applied to the valve.

The unit shown is a 12V unit, and is the kind of thing that you might find in a domestic washing machine. You also often encounter 120V or 220V AC versions of these valves for use in domestic appliances. Having high voltages near water is fraught with danger, so I strongly recommend sticking to 12V pumps and valves for your own projects.

You will also find door latches that are controlled by a solenoid.

Summary

In this chapter, you have learned about how DC motors work, as well as how to switch a motor on and off using an Arduino or Raspberry Pi. You also learned about controlling the speed of a motor.

The next chapter will show you how to use a circuit called an H-bridge to control the direction in which the motor turns.

Advanced Motor Control

8

In the previous chapter, you got as far as controlling the speed of the motor, but not the direction in which it turns. This chapter explores different options for controlling a motor's direction. This includes looking at a number of special-purpose ICs and modules designed to make it easier to control the direction as well as the speed of DC motors.

It is often handy to be able to reverse the direction of a motor. For example, linear actuators that open and close windows or doors work by running a DC motor in one direction to open and then turn the motor in the opposite direction to close. Similarly, if you are making a small robot, you will probably want it to be able to turn its wheels in both directions.

Imagine a motor with two leads, A and B (Figure 8-1).

When A is positive and B is negative, the motor will turn in one direction. If you reverse the polarity of the connections, the motor will turn in the opposite direction.

This means that if you want to control the direction of the motor, you need some way of reversing the polarity of the electricity going to the motors. The way to do this is using a circuit called an H-bridge.

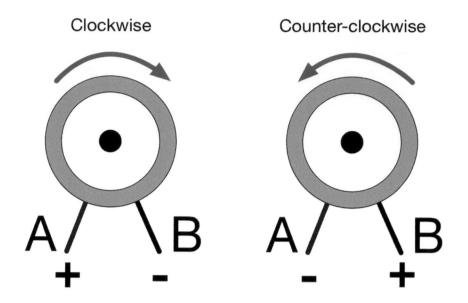

Figure 8-1 *Controlling the direction of a motor*

H-Bridges

Figure 8-1 shows how an H-bridge works. We'll first use switches, before moving on to using transistors or ICs.

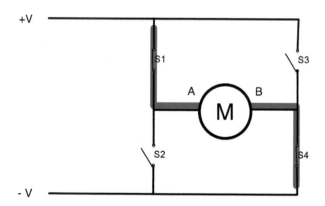

Figure 8-2 *An H-bridge using switches*

If all four switches are open, no current can flow through the motor. However, if S1 and S4 are closed, but S2 and S3 are open (as shown in Figure 8-2), current will flow from the positive supply through terminal A of the motor, through the motor and S4 to the negative supply, and the motor will turn in one direction.

If S1 and S4 are now opened and S3 and S2 closed, the positive supply will this time be applied to terminal B of the motor, and flow out through the motor and S2 to reverse the direction of the motor.

Table 8-1 summarizes how the motor will behave; 0 means the switch is open, 1 means it's closed (conducting), and X means it doesn't matter which state it is in.

Table 8-1 *Switch combinations*

S1	S2	S3	S4	Motor
1	0	0	1	Turns clockwise
0	1	1	0	Turns counterclockwise
0	0	0	0	Motor stopped
1	1	X	X	SHORT CIRCUIT
X	X	1	1	SHORT CIRCUIT
1	0	1	0	Braking
0	1	0	1	Braking

We have already discussed how the H-bridge can be used to switch the direction of the motor, however there are some other combinations of switch settings that you need to know about.

First, and fairly obviously, if all the switches are open, no power will get to the motor and it will quickly come to a stop.

Very importantly, there are some combinations of switches that will directly connect the positive supply to the negative one. This is called a short circuit and is likely to be disastrous as a very large current will flow.

Another situation is where there is no short circuit, but the motor pins are effectively connected together. This has the interesting effect of making the motor brake, slowing faster if it was previously moving and also resisting being turned if it was already stationary. So, if the motor was driving the wheels of a toy rover, using a brake mode might stop the rover rolling away if it was on a slope.

H-Bridge on a Chip

An easy-to-use H-bridge IC that is popular with hobbyists is the L293D. You will use this in "Experiment: Control the Direction and Speed of a Motor" on page 130. This device is great for small motors up to a maximum current of 600mA and a voltage of 36V. You can find out all about this IC from its datasheet (*http://www.ti.com/lit/ds/symlink/l293.pdf*).

The L293D contains two H-bridges plus a bit of extra circuitry to automatically shut down the IC if it starts to get too hot. Although it's still quite possible to destroy an L293D through abuse, you really have to try quite hard to kill it.

The headlines for this chip are:

- Motor voltage range of 4.5 to 36V
- Continuous motor current of 600mA
- Peak motor current of 1.2A
- Diodes on all outputs to protect from motor transient voltage spikes
- Thermal protection
- Compatible with 3V and 5V logic (Pi and Arduino)

Figure 8-3 shows the schematic diagram for this chip and how you could use it to control two DC motors. The chip is actually organized as four half-H-bridges rather than two full H-bridges. You can think of each half-H-bridge as a high-power digital output capable of sourcing or sinking currents up to 600mA. This allows greater flexibility when using the chip.

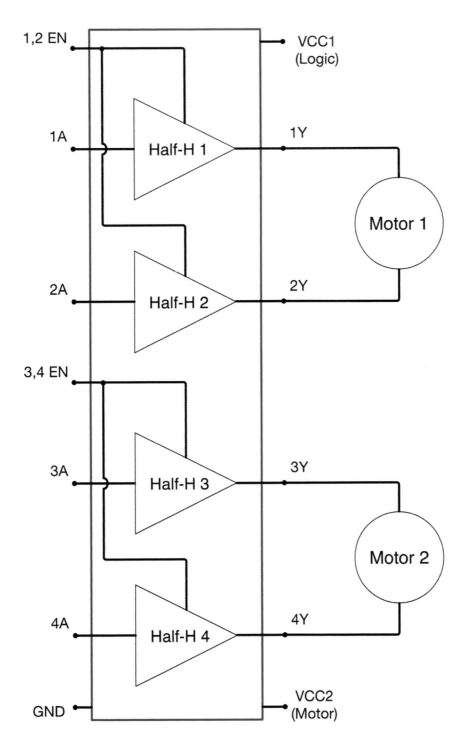

Figure 8-3 *Pinout of the L293D*

The chip has separate pins for the logic and motor power supplies, allowing you to, say, control 6V motors using the 3.3V logic of a Raspberry Pi (as you will see in "Experiment: Control the Direction and Speed of a Motor" on page 130).

The function of each of the pins in Figure 8-3 is described in Table 8-2.

Table 8-2 *L293D pinout*

Pin number	Pin name	Description
1	1,2EN	This pin enables the outputs of half-bridges 1 and 2 (that is, unless this pin is high, those outputs will do nothing); this is often used with a PWM signal to control the overall speed of the motor
2	1A	The input control to half-bridge 1; if this is high, the output on pin 3 will be high (top transistor on)
3	1Y	The output of half-bridge 1
4,5, 12, 13	GND	Ground (on a PCB, all these pins would be soldered to a large pad to act as a heat sink)
6	2Y	The output of half-bridge 2
7	2A	The input control to half-bridge 2
8	VCC2	Voltage supply for the motors, up to 36V; keeping the motor and logic supplies separate helps stability
9	3,4EN	Enables half-bridges 3 and 4
10	3A	The input control to half-bridge 3
11	3Y	The output of half-bridge 3
14	4Y	The output of half-bridge 4
15	4A	The input control to half-bridge 4
16	VCC1	Voltage supply for the logic; this can be lower than the motor voltage supply on pin 8 and will usually be 5V

You will be using this chip in "Experiment: Control the Direction and Speed of a Motor" on page 130 to control both the speed and direction of a DC motor.

Experiment: Control the Direction and Speed of a Motor

In this experiment, you will use an L293D IC on a breadboard. You can see the breadboard connected to a Raspberry Pi in Figure 8-4.

Figure 8-4 *Controlling the speed and direction of a motor using Raspberry Pi*

The hardware for this project is the same for Raspberry Pi and Arduino, and you will be able to experiment with the hardware on its own before connecting it up to your Arduino or Raspberry Pi.

Parts List

Whether you are using a Raspberry Pi or Arduino (or both), you are going to need the following parts to carry out this experiment:

Name	Part	Sources
IC1	L293D H-bridge IC	Adafruit: 807 Mouser: 511-L293D
C1	100nF capacitor	Adafruit: 753 Mouser: 810-FK16X7R2A224K
C2	100µF 16V capacitor	Adafruit: 2193 Sparkfun: COM-00096 Mouser: 647-UST1C101MDD
M1	Small 6V DC motor	Adafruit: 711
	6V (4 x AA) battery box	Adafruit: 830
	400-point solderless breadboard	Adafruit: 64

| Male-to-male jumper wires | Adafruit: 758 |
| Female-to-male jumper wires (Pi only) | Adafruit: 826 |

The project will work without the capacitors C1 and C2 just for this experiment, but at some point you will need to get into the good habit of using capacitors if this becomes more than just an experiment and you want to deploy the project for real.

If you are planning to try this experiment with a Raspberry Pi, you will need female-to-male jumper wires to connect the Raspberry Pi GPIO pins to the breadboard.

Design

Figure 8-5 shows the schematic diagram for this project.

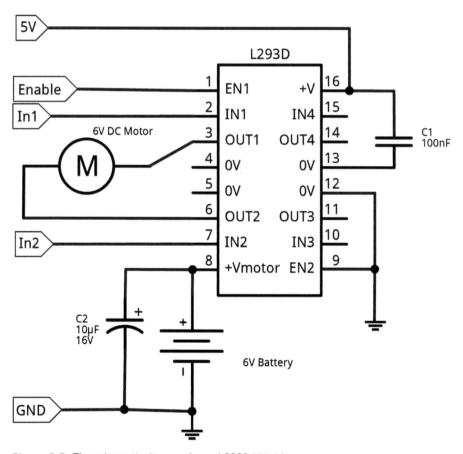

Figure 8-5 *The schematic diagram for an L293D H-bridge*

The Raspberry Pi or Arduino provide the 5V logic supply to pin 16 and the power for the motor is supplied to pin 8 from a 6V battery pack.

Only one of the H-bridges in the IC is actually used and so the EN2 pin is connected to ground to disable the unused half of the chip.

The EN1, IN1, and IN2 pins will all be connected to digital output pins on the Raspberry Pi or Arduino.

Capacitors

The capacitors are only optional, because if we are just experimenting with this circuit for a few hours, the reliability that the capacitors add is not really a problem.

This arrangement of capacitors is very typical for an H-bridge IC circuit. C1 is called a decoupling capacitor. It should be positioned as closely as possible to the IC and be between the logic sup-

ply and GND. This can be just 100nF (a small value capacitor) and it removes any electrical noise that might interfere with the operation of the chip's logic.

C2 provides a short-term reservoir of energy, but for the motors rather than the switching logic. The value of this capacitor is usually much larger than C1, typically 100μF or more.

Breadboard Layout

Before connecting the H-bridge to an Arduino or Raspberry Pi, you can experiment with it on its own, supplying both the motor and logic power from the same 6V battery pack. This is fine for experimenting with the chip on its own, but when you come to using it with an Arduino or Raspberry Pi, you will separate the supplies, so that the motor is supplied by the battery pack and the IC's logic supply comes from the Arduino or Raspberry Pi.

Figure 8-6 shows the breadboard layout for standalone experimentation. The only thing that changes when you come to connect it to an Arduino or Raspberry Pi is some of the jumper wires.

Wire up the breadboard paying particular attention to the IC. You need to make sure to position it correctly—the little notch on one end of the IC should be toward the top of the breadboard on row 10. The pin to the top left of the notch is pin 1.

You can also connect up the battery. Initially the motor should not be turning.

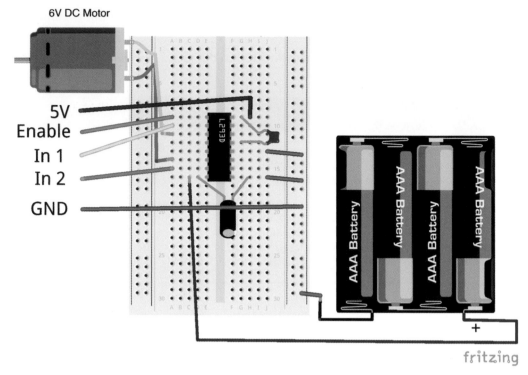

Figure 8-6 *The breadboard layout for standalone testing of the H-bridge*

 ## Clockwise and Counterclockwise

Although I talk about the motor turning clockwise or counterclockwise, that depends on the motor, so if your motor turns counterclockwise when I say it should turn clockwise, that's not a problem as long as it turns clockwise when I say it should turn counterclockwise.

You can also swap the motor leads over, if you want the rotation to match the description.

It can be difficult to tell which way the motor is turning, if it's just a bare metal shaft. The sense of touch is a wonderful thing, so if you just gently pinch the motor shaft between thumb and forefinger, you will be able to feel which way it's turning.

Another technique is to cut a very short length of painter's tape and stick it over the spindle so that it acts as a kind of flag.

Experimenting

Whether we want the motor to turn clockwise or counterclockwise, the Enable pin needs to be connected to +V. To make the motor turn in one direction, connect the other end of the jumper wire connected to In 1 to +V and connect In 2 to the ground column on the right of the breadboard (Figure 8-7). Note that in Figures 8-7 and 8-8, the IN1 and IN2 jumper wires are made thicker to distinguish them from the other jumper wires.

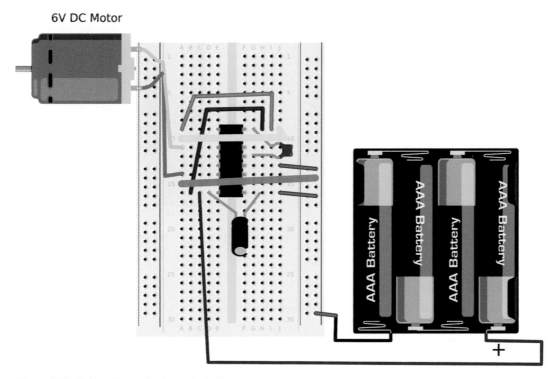

Figure 8-7 *Making the motor turn clockwise*

Next, you are going to reverse the direction of the motor. The connection to Enable can stay where it is, but you need to reverse In 1 and In 2, so that In 1 is now connected to ground and In 2 to +V, as shown in Figure 8-8.

Now that we know that the H-bridge is working on its own, you can use it with an Arduino (or, if you prefer to use the Raspberry Pi, you can skip ahead to "Connecting the Raspberry Pi" on page 140).

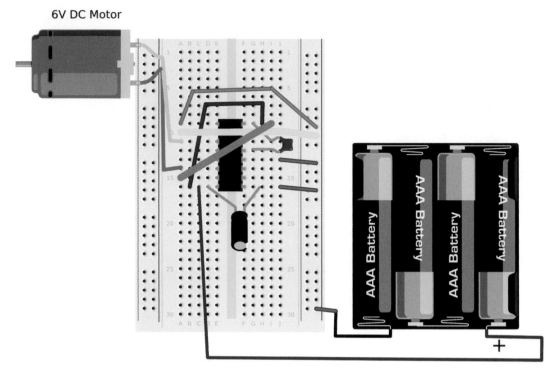

Figure 8-8 *Making the motor turn counterclockwise*

Arduino Connections

Figure 8-9 shows how you should connect the breadboard to an Arduino Uno using jumper wires.

Arduino pin 11 is used for Enable, because this Arduino is capable of PWM and we will use the Enable pin of the L293D to control the speed of the motor. The pins connected to In 1 and In 2 could be any Arduino digital pins; 10 and 9 were chosen for no better reason than that they are close to 11 and it's neater with all the wires together.

The Arduino provides 5V to the logic side of the L293D, but importantly, this is not supplying power to the motor. The motor power still comes from the battery pack.

Figure 8-10 shows the breadboard connected to the Arduino with male-to-male jumper wires and ready to go.

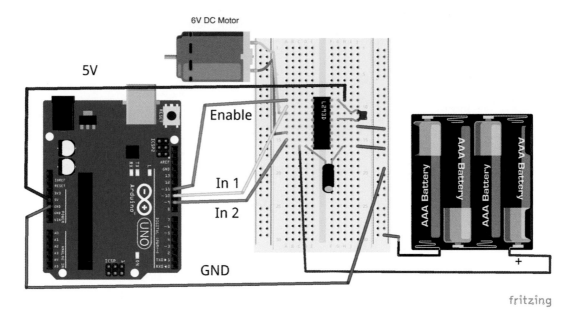

Figure 8-9 *Connecting the Arduino and H-bridge*

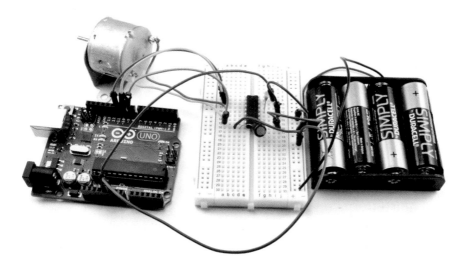

Figure 8-10 *Full control of a motor with Arduino*

Arduino Software

The Arduino software that you used back in "Experiment: Controlling the Speed of a DC Motor" on page 96 (PWM control of a motor) uses the Serial Monitor to allow you to send speed commands to the motor. In this experiment, this is extended, so that as well as sending a speed command, a direction command is also sent.

The Arduino sketch is in */arduino/experiments/full_motor_control* (which you'll find in the place where you downloaded the book's code—see "The Book Code" on page 14 in Chapter 2):

```
const int enablePin = 11;    ❶
const int in1Pin = 10;
const int in2Pin = 9;

void setup() {               ❷
  pinMode(enablePin, OUTPUT);
  pinMode(in1Pin, OUTPUT);
  pinMode(in2Pin, OUTPUT);
  Serial.begin(9600);
  Serial.println("Enter s (stop) or f or r followed by Duty Cycle (0 to 255). E.g.
f120");
}

void loop() {                        ❸
  if (Serial.available()) {
    char direction = Serial.read();  ❹
    if (direction == 's') {          ❺
      stop();                        ❻
      return;
    }
    int pwm = Serial.parseInt();     ❼
    if (direction == 'f') {          ❽
      forward(pwm);
    }
    else if (direction == 'r') {
      reverse(pwm);
    }
  }
}

void forward(int pwm)                ❾
{
  digitalWrite(in1Pin, HIGH);
  digitalWrite(in2Pin, LOW);
  analogWrite(enablePin, pwm);
  Serial.print("Forward ");
  Serial.println(pwm);
}

void reverse(int pwm)                ❿
{
  digitalWrite(in1Pin, LOW);
  digitalWrite(in2Pin, HIGH);
  analogWrite(enablePin, pwm);
  Serial.print("Reverse ");
  Serial.println(pwm);
}

void stop()                          ⓫
```

```
{
  digitalWrite(in1Pin, LOW);
  digitalWrite(in2Pin, LOW);
  analogWrite(enablePin, 0);
  Serial.println("Stop");
}
```

This sketch is relatively long, but is well structured into functions that will form a good basis for changing the code or reusing it in your projects.

❶   The sketch starts by defining constants for the three control pins.

❷   The setup() function sets these pins to be outputs and then starts serial communication at 9600 baud and sends an instruction message to the Serial Monitor, reminding you of the format of the messages to send to control the motor.

❸   The loop() function does nothing unless Serial.available reports that there is a new incoming message from the Serial Monitor.

❹   The first character of this message is read as the direction letter, which can either be s (stop), f (forward), or r (reverse).

❺   If the direction letter is *s* then the stop function is called and the return command ensures that none of the subsequent code in the loop function is run.

❻   The *s* command does not need a pwm parameter, so we don't want to try and read it from the message because it won't be there.

❼   If the direction code is f or r (not *s*), the value of pwm is taken from the rest of the message using parseInt.

❽   One of the functions forward or reverse is then called depending on the direction character.

❾   The forward function sets in1Pin to HIGH and in2Pin to LOW to set the direction of the motor and then uses analogWrite to control the speed of the motor using the enable Pin and the pwm parameter to forward. Finally, a message confiming the action is sent back to the Serial Monitor.

❿   The reverse function is almost the same as forward except that in1Pin is set to LOW and in2Pin to HIGH to make the motor spin in the opposite direction.

⓫   The stop function sets all the control pins to be LOW.

Arduino Experimentation

To try out the experiment with Arduino, connect the USB lead to your Arduino and upload the sketch. Open the Serial Monitor (Figure 8-11) type the command *f100* in the top of the

Serial Monitor and click Send. The motor should start running in one direction at a fairly slow speed.

Figure 8-11 *Controlling a motor with serial commands*

Next, try entering the command *f255*. This will set the motor running at full speed. The *s* command will stop the motor and *r100* will set it running slowly in the opposite direction. *r255* will set it running in reverse at full speed.

Connecting the Raspberry Pi

An advantage of using an H-bridge IC like the L293D is that the control pins require very little current to control the motor. In fact, the datasheet specifies that this will always be less than 100µA (0.1mA), which means there is no problem at all using the Raspberry Pi's low current outputs.

If you have both a Raspberry Pi and an Arduino, and you have just finished the Arduino part of this experiment, then to get the breadboard set up and ready for the Raspberry Pi, you just need to swap out the male-to-male jumper wires that were connecting the breadboard to the Arduino with female-to-male jumpers to connect the breadboard to the GPIO pins of your Raspberry Pi.

Figure 8-12 shows how your Raspberry Pi needs to be connected to the breadboard. Figure 8-4 is a photograph of the same setup.

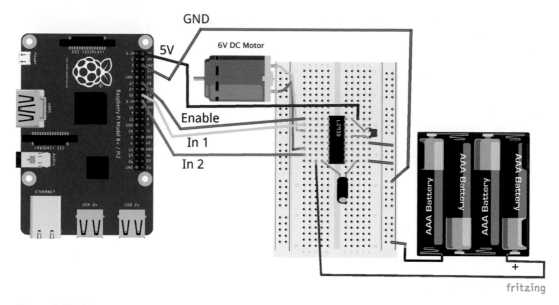

Figure 8-12 *Connecting the Raspberry Pi and H-bridge*

Raspberry Pi Software

You will find the code for the following program in the *full_motor_control.py* file, which is in the *python/experiments* directory (you'll find this in the place where you downloaded the book's code):

```python
import RPi.GPIO as GPIO
import time

GPIO.setmode(GPIO.BCM)

enable_pin = 18      # ❶
in_1_pin = 23
in_2_pin = 24

GPIO.setup(enable_pin, GPIO.OUT)
GPIO.setup(in_1_pin, GPIO.OUT)
GPIO.setup(in_2_pin, GPIO.OUT)
motor_pwm = GPIO.PWM(enable_pin, 500)
motor_pwm.start(0)

def forward(duty):           # ❷
    GPIO.output(in_1_pin, True)
    GPIO.output(in_2_pin, False)
    motor_pwm.ChangeDutyCycle(duty)

def reverse(duty):           # ❸
    GPIO.output(in_1_pin, False)
    GPIO.output(in_2_pin, True)
    motor_pwm.ChangeDutyCycle(duty)
```

```
def stop():
    GPIO.output(in_1_pin, False)
    GPIO.output(in_2_pin, False)
    motor_pwm.ChangeDutyCycle(0)

try:
    while True:            # ❹
        direction = raw_input('Enter direction letter (f - forward, r - reverse, s -
stop): ')
        if direction[0] == 's':
            stop()
        else:
            duty = input('Enter Duty Cycle (0 to 100): ')
            if direction[0] == 'f':
                forward(duty)
            elif direction[0] == 'r':
                reverse(duty)

finally:
    print("Cleaning up")
    GPIO.cleanup()
```

This code borrows a lot from the earlier Python code for experiments.

❶   At the top of the file is the usual GPIO setup code and pin definitions. The Enable pin of the L293D is used to control the speed of the motor, so pin 18 that is connected to it is configured as a PWM output.

❷   The forward function sets the IN1 and IN2 pins to control the direction and then sets the duty of the PWM channel.

❸   If you compare it with the reverse function, you can see that the values of the IN1 and IN2 pins are swapped over. The stop function sets the direction pins to the stop position (both LOW) and the duty cycle to 0.

❹   The main while loop prompts the user for a command and then calls stop, forward, or reverse, as appropriate.

Raspberry Pi Experimentation

The program for experimenting with Raspberry Pi is called *ex_04_full_motor_control.py*. Run it and you will see that it operates in much the same way as its Arduino counterpart, prompting you for direction and speed.

```
$ sudo python ex_04_full_motor_control.py
Enter direction letter (f - forward, r - reverse, s - stop): f
Enter Duty Cycle (0 to 100): 50
Enter direction letter (f - forward, r - reverse, s - stop): f
Enter Duty Cycle (0 to 100): 100
Enter direction letter (f - forward, r - reverse, s - stop): s
Enter direction letter (f - forward, r - reverse, s - stop): r
```

```
Enter Duty Cycle (0 to 100): 50
Enter direction letter (f - forward, r - reverse, s - stop): r
Enter Duty Cycle (0 to 100): 100
Enter direction letter (f - forward, r - reverse, s - stop): s
Enter direction letter (f - forward, r - reverse, s - stop):
```

Be Nice to Your Motor

Imagine a car driving along and then suddenly being thrown into reverse gear—that is pretty much what you are doing if you suddenly reverse the direction of a motor. For small motors, without a great deal of mass attached to them, this isn't normally much of a problem. You may find that if you are using a Raspberry Pi or Arduino that is powered from the same power source as the motors, then the Raspberry Pi may crash or the Arduino may reset. This happens as a result of the large current that flows when you suddenly switch directions, causing the power supply voltage to dip.

For larger motors, sudden changes in speed or direction to something that has a lot of inertia can cause big problems. As well as the resultant large currrents that will flow and may damage the H-bridge, there is also the mechanical shock to the bearings of the motor.

This is something to bear in mind when designing the control software that uses larger motors. One way to be nicer to your motors is to precede any change in direction by setting the control lines to let the motor stop, pausing for enough time for it to actually stop before setting it running again in the opposite direction.

In Arduino, assuming that you have the utility functions forward and reverse from "Experiment: Control the Direction and Speed of a Motor" on page 130, this might look something like this:

```
forward(255);
delay(200);
reverse(255);
```

Other H-Bridge ICs

There are many motor controller ICs on the market. In this section, we will explore some of the more commonly used devices.

L298N

At 600mA, the maximum current of the L293D is quite low. If you need a bit more, then fortunately the L293D has a bigger brother in the L298N (*https://www.sparkfun.com/datasheets/Components/General/L298N.pdf*).

The headlines for this chip are:

- Motor voltage range of up to 50V

- Continuous motor current of 2A per motor

- Peak motor current of 3A per motor

- Compatible with 3V and 5V logic (Pi and Arduino)

Figure 8-13 shows the pinout for the device. The configuration is very similar to the L298D, but the L298D also adds the ability to monitor the current flowing through the motor if you add a couple of low-value resistors.

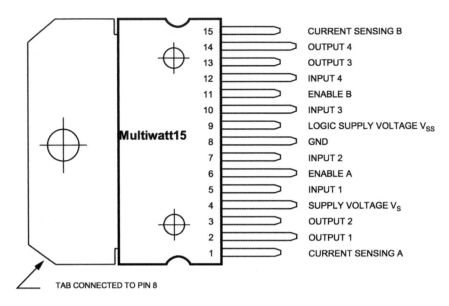

15	CURRENT SENSING B
14	OUTPUT 4
13	OUTPUT 3
12	INPUT 4
11	ENABLE B
10	INPUT 3
9	LOGIC SUPPLY VOLTAGE V_{SS}
8	GND
7	INPUT 2
6	ENABLE A
5	INPUT 1
4	SUPPLY VOLTAGE V_S
3	OUTPUT 2
2	OUTPUT 1
1	CURRENT SENSING A

Multiwatt15

TAB CONNECTED TO PIN 8

Figure 8-13 *The L298N Dual H-bridge IC*

As you can see from Figure 8-13, the IC is in a package designed to be bolted to a heat sink for higher power switching. The pins are arranged in two rows and offset, which means they do not fit onto a standard breadboard. The easiest way to use the device is to buy a low-cost, ready-made H-bridge module that uses it. These can be bought for a few dollars (see "H-Bridge Modules" on page 147). The pins of the L298N are listed in Table 8-3.

Table 8-3 *L298N pinout*

Pin number	Pin name	Description
1	CURRENT SENSE A	See the discussion after this table
2	OUTPUT 1	The output of half-bridge 1
3	OUTPUT 2	The output of half-bridge 2
4	Vs	Voltage supply for the motors
5	INPUT 1	The input control to half-bridge 1
6	ENABLE A	Enables half-bridges 1 and 2
7	INPUT 2	The input control to half-bridge 2
8	GND	Ground

9	Vss	Voltage supply for the logic; this can be lower than the motor voltage supply on pin 8 and will usually be 5V
10	INPUT 3	The input control to half-bridge 2
11	ENABLE B	Enables half-bridges 1 and 2
12	INPUT 4	The input control to half-bridge 4
13	OUTPUT 3	The output of half-bridge 3
14	OUTPUT 4	The output of half-bridge 4
15	CURRENT SENSE B	See the discussion after this table

Most of the pins have direct counterparts on the L293D. Just like the L293D, the half-bridges can be enabled in pairs, to allow PWM control of the power to the motor when used as an H-bridge.

Being able to measure the current flowing to the motors can be useful if, for example, you want to detect that a motor is not moving, because something is preventing it from turning. When this happens, the current in the motor greatly increases. This is known as the stall current of the motor. If you do not need to measure the current flowing through each of the two H-bridges, then pins 1 and 15 must be connected to GND.

To measure the current, you attach low-resistance, but high-power rating resistors between pin 1 and GND and also between pin 15 and GND. The resistors need to be low value so that they do not noticeably interfere with the motor's operation and also high enough power to cope with the heat that they will generate. The voltage across the resistor will be proportional to the current flowing. This voltage could then be measured by an Arduino analog input (see "Analog Inputs" on page 18 in Chapter 2).

As an example, let's say you are driving a 12V motor that normally uses 500mA but when stalled uses a current of 2A. You can easily afford to sacrifice up to 0.5V to be able to detect the motor stalling.

Figure 8-14 shows what's going on with the L298N, the resistor, and motor, for one of the two H-bridges.

The CURRENT SENSE A pin is connected to an Arduino analog input (let's say A0), and the two pins IN1 and IN2 are connected to Arduino digital outputs D2 and D3 to control the direction of the motor.

The maximum current is 2A, and you are aiming for 0.5V across the resistor. To get that, we would need to use a resistor of value R = V/I = 0.5/2 = 0.25Ω. Converting this to the nearest standard resistance value, that would be 0.27Ω or 270mΩ (milli-ohms—not to be confused with Mega-Ohms). This means that at 2A, the actual voltage would be V = I x R = 2A x 0.27Ω = 0.54V.

To calculate the power rating of the resistor, multiply the voltage (0.54V) by the current (2A) to give a power rating of a little over 1W. You would probably use a 2W resistor to allow some margin of error.

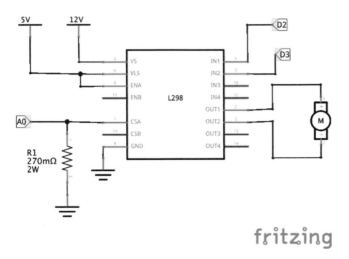

Figure 8-14 *Measuring the motor current*

The CURRENT SENSE A pin can be connected directly to an Arduino analog input (the Raspberry Pi does not have analog inputs) and the voltage measured. Because the voltage is lower than the full 5V range that the Arduino analog input can use, the raw `analogInput` reading at 2A (0.5V) will be around 100, still giving us a reasonable degree of accuracy.

The following Arduino sketch shows how you might have the Arduino automatically turn off the motors if the current exceeded 1.5A, indicating the motor stalling:

```
const float R = 0.27;
const int in1pin = 2;
const int in2pin = 3;
const int sensePin = A0;

void setup() {
  pinMode(in1pin, OUTPUT);
  pinMode(in2pin, OUTPUT);
  // set motor to go forwards
  digitalWrite(in1pin, HIGH);
  digitalWrite(in2pin, LOW);
}

void loop() {
  int raw = analogRead(sensePin);
  float v = raw / 204.6; // 204.6 = 1024 / 5V
  float i = v / R;
  if (i > 1.5) {
    // stop the motor
    digitalWrite(in1pin, LOW);
```

```
        digitalWrite(in2pin, LOW);
    }
}
```

The loop function takes an analog reading. This reading from an analog input has a maximum value of 1023 for 5V, so to turn the raw analog reading into a voltage it needs to be divided by 204.6, which is 1023 / 5.

The current can then be calculated using Ohm's law.

TB6612FNG

Both the L293D and L298N have been around for many years and are somewhat old-fashioned devices that use bipolar transistors rather than MOSFETs and therefore tend to get hot even for relatively low currents.

A much more modern device that is similar in power rating to the L293D is the TB6612FNG (*https://www.sparkfun.com/datasheets/Robotics/TB6612FNG.pdf*). The TB6612FNG is also a dual H-bridge device.

It only has a 15V maximum motor voltage and the combined current from both H-bridges needs to stay under 1.2A with a peak of 3.2A. The device also has built-in thermal shutdown.

This device is only available in a surface-mount chip package, but you can buy ready-made modules that use this IC and will fit on the breadboard.

H-Bridge Modules

You can avoid making your own H-bridge by using an H-bridge module. These come in all sorts of shapes and sizes to suit different motor currents. Figure 8-15 shows a selection of H-bridge motor controller modules.

Figure 8-15 *H-bridge modules*

The module on the left of Figure 8-15 is a very low-cost module from eBay that uses two L9110S ICs, each IC being a single H-bridge. These modules are available on eBay for a couple of dollars. The four screw terminals connect to two DC motors and the header pins are just GND, VSS (5V low current), and then four direction control pins, just like the L293D.

The module in the center of Figure 8-15 uses a TB6612FNG and is made by Sparkfun (product code ROB-09457). You can solder your own header pins onto the connectors to attach it to the breadboard, or as shown here, header sockets to use male-to-male jumper wires to connect it up.

On the right of Figure 8-15 is a module that uses the L298N, complete with heat sink. The module also includes protection diodes down each side of the heat sink and even includes a voltage regulator that can supply 5V power to your Arduino, but probably not enough current for a Raspberry Pi. It does not have current sensing resistors.

Even higher power motor controllers are available and a good place to find them is on the Pololu website (*https://www.pololu.com*). Here you will find H-bridges that can cope with tens of amps. As the current increases, then so of course does the price.

As well as separate H-bridge modules, you can also find shields that plug onto an Arduino and plug-on boards that fit onto a Raspberry Pi. Figure 8-16 shows three such boards.

Figure 8-16 *H-bridge shields for Arduino and Raspberry Pi*

The board on the left of Figure 8-16 is an Arduino motor shield from Sparkfun (product code DEV-09815) based on the L298P a surface mount version of the L298N. This board has a prototyping area to which other components can be attached. The board in the center is a high-power H-bridge shield for the Arduino sold by Polulu that can very impressively switch 30A loads.

On the right of Figure 8-16 is the RasPiRobot V3 board that fits onto the GPIO headers of a Raspberry Pi and uses a TB6612FNG to control two DC motors.

Project: Arduino Beverage Can Crusher

Remember the linear actuator from "Linear Actuators" on page 120 in Chapter 7? Well, you can put one of these together with an H-bridge module and an Arduino, and with a bit of woodwork, make yourself an automatic beverage can crusher (Figure 8-17).

Figure 8-17 *An Arduino can crusher*

You can see a video of a prototype of the crusher in action at https://youtu.be/qbWMEIFnq2I.

 Linear Actuators Are Strong

This project is about crushing beverage cans, but the actuator will quite happily crush a hand or anything else that you are foolish enough to put into the crushing area. So, be careful, especially while you are getting it to work, where there is a temptation to intervene.

Parts List

In addition to an Arduino Uno, you need the following parts to build this project:

Part	Sources
12V 6-inch linear actuator	eBay
L298 H-bridge module	eBay
2 x female-to-male jumper wires	Adafruit: 826
4 x male-to-male jumper wires	Adafruit: 758

Female barrel jack to screw terminal adapter	Adafruit: 368
Power supply (12V at 3A minimum)	Adafruit: 352
Lots of 2x4-inch wood and a bit of plywood	Hardware store
Wood screws and woodworking tools	Hardware store

Linear actuators vary considerably in price. For can crushing, you need something with a 6-inch travel. Check the maximum current of the linear actuator that you select and choose an H-bridge module that will cope. The linear actuator that I used has a maximum current of 3A so I used an H-bridge based on the L298 chip.

Wiring

Figure 8-18 illustrates the wiring diagram for the project, and Figure 8-19 shows a close-up of the Arduino and H-bridge module.

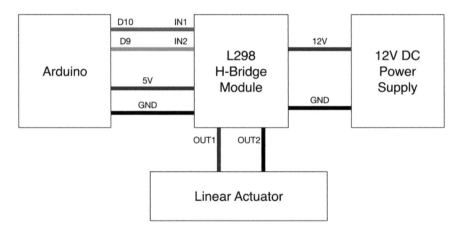

Figure 8-18 *Wiring diagram for the can crusher*

The L293 module has jumper pins that by default keep both H-bridges enabled, and so only the two Arduino outputs connected to IN1 and IN2 are needed.

Conveniently, the H-bridge module included a voltage regulator that provides a 5V output that can be connected to the 5V pin of the Arduino to supply it with power.

Figure 8-19 *The Arduino and H-bridge module*

Mechanical Construction

As you can see from Figure 8-17, the basic structure of the project is a length of 2x4 wood. At one end of this, the actuator is anchored using the fixtures supplied with it. The actuator's shaft is then fixed to a block of wood that crushes the can against an end stop. Two plywood sides help prevent the can escaping while it's being crushed.

I have not used exact dimensions, because your actuator is likely to be a little different in size than mine. The best way to get it right is to place the actuator on the wood and then calculate the spacings. Remember to leave a bit of a gap between the fully extended crushing surface and the end stop; otherwise, the machine might push itself apart.

Arduino Software

To Reset button of the Arduino is used to trigger the crushing action. Whenever the Arduino resets, it automatically starts to move the linear actuator. You can find the code for this project in the *pr_02_can_crusher* file:

```
const int in1Pin = 10;
const int in2Pin = 9;

const long crushTime = 30000;  // ❶

void setup() {     // ❷
  pinMode(in1Pin, OUTPUT);
  pinMode(in2Pin, OUTPUT);
```

```
    crush();
    stop();
    delay(1000);
    reverse();
    stop();
  }

  void loop() {
  }

  void crush()
  {
    digitalWrite(in1Pin, LOW);
    digitalWrite(in2Pin, HIGH);
    delay(crushTime);
  }

  void reverse()
  {
    digitalWrite(in1Pin, HIGH);
    digitalWrite(in2Pin, LOW);
    delay(crushTime);
  }

  void stop()
  {
    digitalWrite(in1Pin, LOW);
    digitalWrite(in2Pin, LOW);
  }
```

❶   Although the actuator will automatically stop when it gets to the end of its travel, this period (of 30 seconds for my motor) sets how long the motor should be on for, before reversing, ready for the next can.

❷   The setup function controls the whole operation of the project. After setting both control pins to be outputs, it immediately starts the crushing action using the crush function.

Summary

In this chapter, you have learned how to control the direction of DC motors using an H-bridge. In fact, H-bridges can be used to control other types of motors, including stepper motors (discussed in Chapter 10). H-bridges can also be used to switch power to other devices, such as Peltier heating and cooling elements (discussed in Chapter 11).

Servomotors

## 9

Servomotors (not to be confused with stepper motors, which we will discuss in Chapter 10) combine a small DC motor, some gears, and a control circuit including a variable resistor to be able to position the arm of a servo at a certain angle.

This makes them really handy for projects where you want to make something move fairly quickly and relatively precisely.

Servomotors

While you can get continuous servos that can rotate continuously, most servomotors (or just servos) can only rotate through about 180°. They are often used in remote control vehicles to control steering, or the angles of surfaces on an R/C airplane. Figure 9-1 shows two different sizes of servo.

Figure 9-1 *A 9g and standard servomotor*

The servo on the right is a so-called *standard* servo. This is the most common size of servo and they are indeed pretty standard, even often having the same mounting hole spacings and sizes. The much smaller servo on the left is intended for flying vehicles and is light-weight. These servos are called 9g servos.

The servos shown in Figure 9-1 are often called hobby servos. They vary in quality, with the higher-quality, higher-torque servos having metal rather than nylon gears. Most servos operate on around 5V, often with an acceptable voltage range of around 4 to 7V. Hobby servos have leads terminating in a three-way socket (+power, -power, and control signal).

Larger, sometimes much larger servomotors are also available for high-power applications, but these are not as standardized as hobby servos.

How a Servomotor Works

Figure 9-2 shows how a servomotor works.

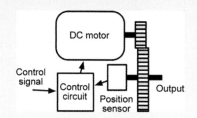

Figure 9-2 *How a servomotor works*

A servo is made up of a DC motor that drives a reduction gearbox to reduce the rotational speed of the motor while at the same time increasing the torque. The output drive shaft is connected to a position sensor (usually a variable resistor) to monitor the position of the shaft. A control circuit uses this input from the position sensor combined with a control signal that sets the desired position and uses that to control the power and direction that the DC motor turns.

The control unit takes the desired position and subtracts the actual position to give the "error," which may be positive or negative. This error is then used to power the motor. The larger the difference between the desired and actual positions of the output shaft, the faster the motor will turn toward the desired position. As it gets closer to zero error, the power to the motor decreases.

Controlling a Servo

The control signal to the servo is not, as you might expect, a voltage, but rather it is a PWM signal. This signal is standard across all hobby servos and looks like Figure 9-3.

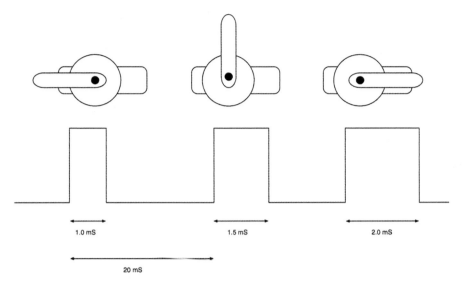

Figure 9-3 *Controlling a servomotor*

A train of pulse of 1.5 milliseconds will set the servo to its center 90° position. Shorter pulses of 1.0 milliseconds will set it to 0° and 2.0 milliseconds to 180°. In reality, this range may be a little less than the full 180°, without making the pulses shorter at one end and longer at the other. In fact, it is not uncommon for the 0° pulse to need to be 0.5 milliseconds and the 180° pulse to need to be 2.5 milliseconds.

The servo will expect to see a pulse every 20 milliseconds.

Experiment: Control the Position of a Servomotor

In this experiment, you will use a Raspberry Pi and Arduino to set the position of a servomotor's arm to a particular angle.

The Arduino has a servo library that will generate pulses on any of the pins, so you do not need to use a pin that is marked as PWM capable.

To test out the servo, you will use the Serial Monitor to send the angle you want to set the servomotor to.

Generating accurately timed pulses on a Raspberry Pi is much harder than on an Arduino. An Arduino has hardware that generates the pulses, but a Raspberry Pi generates these pulses using software. Because the Raspberry Pi has an operating system that allows lots of processes to compete for the time of the processor, PWM pulses will sometimes be longer than expected. This leads to the servo sometimes being a bit jittery. Although it is usable as is, if greater accuracy is needed, then you can use external PWM hardware, as described in "Project: Pepe, the Dancing Raspberry Pi Puppet" on page 162.

Hardware

The neat thing about servomotors is that the control electronics for the motor are all included in the motor package, so there is no need for separate H-bridge or transistor drivers for the motor. All that is needed is to supply 5 or 6V to the power supply pins and send low-current pulses from a digital output.

Figure 9-4 shows a servo connected to a Raspberry Pi.

Figure 9-4 *A servomotor controlled by a Raspberry Pi*

You can see from Figure 9-4 that one of the arms (supplied with the servo in a little bag) has been attached to the servo, so you can see the position that the motor is in.

Parts List

Whether you are using a Raspberry Pi or Arduino (or both), you are going to need the following parts to carry out this experiment:

Part	Sources
9g mini servomotor	eBay, Adafruit: 169
Male-to-male jumper wires	Adafruit: 758
Female-to-male jumper wires (Pi only)	Adafruit: 826

If you are planning to try this experiment with a Raspberry Pi, you will need female-to-male jumper wires to connect the Raspberry Pi GPIO pins to the servo.

The small 9g servo should work just fine drawing its power from the 5V supply of the Pi or Arduino, but if you want to use a bigger servo, you might find that you need to use a separate power supply such as the 6V battery pack that you have used in other experiments.

Connecting the Arduino

Figure 9-5 shows how the Arduino and servomotor are wired together.

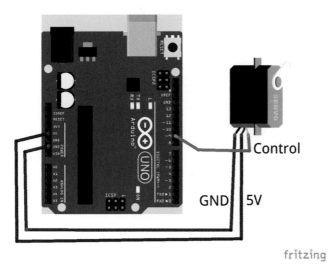

Figure 9-5 *Wiring a servomotor to an Arduino*

Male-to-male jumper wires are used to connect the three-way socket of the servomotor to the Arduino.

Servomotor Connections
Figure 9-6 shows a close-up of a servomotor and its connector. As with many things about servos, these connections are pretty standard across all the servomotor manufacturers.   The connections to the motor are identified by the colors of the wires attached to them. They are almost always in the same order: • Brown (occasionally black) for Ground   • Red for +V   • Orange (occasionally yellow) for Control

Figure 9-6 *Servomotor connection*

Arduino Software

The Arduino servo library is included with the Arduino IDE and takes all the hard work out of working with servos.

The Arduino sketch is in *arduino/experiments/servo* (you'll find this in the place where you downloaded the book's code—see "The Book Code" on page 14 in Chapter 2):

```
#include <Servo.h>

const int servoPin = 9;    // ❶

Servo servo;               // ❷

void setup() {
  servo.attach(servoPin);       // ❸
  servo.write(90);       // ❹
  Serial.begin(9600);           // ❺
  Serial.println("Enter angle in degrees");
}
```

```
void loop() {          // ❻
  if (Serial.available()) {
    int angle = Serial.parseInt();
    servo.write(angle);        // ❼
  }
}
```

Although you could use PWM and analogWrite to generate a pulse of the right duration to control a servo, this would also involve changing the PWM frequency and restrict the pin options for controlling the servo to just the pins that can do PWM. This is the approach you will see in "Raspberry Pi Software" on page 160 but on an Arduino, it is easier and better to use the servo library.

❶   After importing the servo library, a constant (servoPin) is defined as the pin to be connected to the control terminal of the servo.

❷   The variable servo of type Servo is declared. This variable is then used whenever the servo position needs to change.

❸   The setup() function starts the servo pulses by associating the control pin with the pulse generation using servo.attach.

❹   Set the servo angle to 90° (the center position).

❺   setup also starts serial communication so that you can send it angle commands over the Serial Monitor.

❻   The loop function waits for an incoming angle command and then converts it into a number using parseInt.

❼   It then sets the new servo angle using servo.write.

Experimenting with Arduino

Upload the sketch to the Arduino. The motor arm should immediatley spring over to its center position (90°). Now open the Serial Monitor (Figure 9-7) and try entering some angles between 0° and 180°. The motor should snap to its new position each time you enter a new angle.

Figure 9-7 *Controlling the servo's position using the Serial Monitor*

Connecting the Raspberry Pi

To connect up the Raspberry Pi, use female-to-male jumper wires to link the GPIO header pins to the servo, as shown in Figure 9-8.

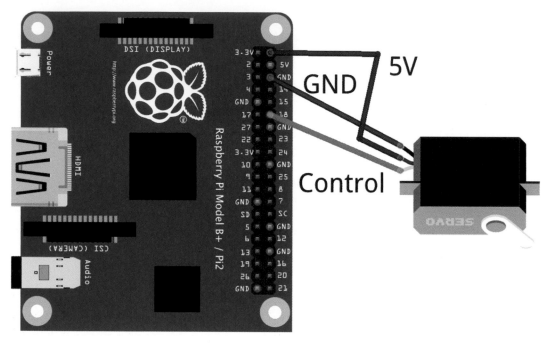

Figure 9-8 *Connecting a Raspberry Pi to a servo*

Raspberry Pi Software

The Python program for this servo control experiment can be found in the file *python/experiments/servo.py*. For help on installing the Python programs on your Raspberry Pi, see "The Book Code" on page 34 in Chapter 3.

The program uses the PWM feature of the RPi.GPIO library to generate the control pulses for the servo.

There are quite a lot of calculations in there to create the correct pulse lengths. It does not matter if you don't follow these, you can just copy the code if you need it for your own projects and just call set_angle to set the angle of the servo arm:

```
import RPi.GPIO as GPIO
import time

servo_pin = 18

# Tweak these values to get full range of servo movement
deg_0_pulse = 0.5   # ms ❶
deg_180_pulse = 2.5 # ms
f = 50.0   #50Hz = 20ms between pulses ❷
```

```
# Do some calculations on the pulse width parameters
period = 1000 / f  # 20ms    ❸
k = 100 / period         # duty 0..100 over 20ms  ❹
deg_0_duty = deg_0_pulse * k   ❺
pulse_range = deg_180_pulse - deg_0_pulse
duty_range = pulse_range * k    ❻

# Initialize the GPIO pin
GPIO.setmode(GPIO.BCM)
GPIO.setup(servo_pin, GPIO.OUT)    ❼
pwm = GPIO.PWM(servo_pin, f)
pwm.start(0)

def set_angle(angle):     ❽
    duty = deg_0_duty + (angle / 180.0) * duty_range
    pwm.ChangeDutyCycle(duty)

try:
    while True:      ❾
        angle = input("Enter angle (0 to 180): ")
        set_angle(angle)
finally:
    print("Cleaning up")
    GPIO.cleanup()
```

❶   Because every servo needs slightly different pulse lengths to maximize their range of angles, two constants (deg_0_pulse and deg_180_pulse) are used to set the pulse durations for an angle of 0° and 180°, respectively. You can tweak these parameters if you need to to get the full range of movement.

The next block of code does some calculations relating to the pulse lengths.

❷   A pulse every 20 milliseconds means we need to set the PWM frequency (f) to be 50 pulses per second.

❸   The period (20 milliseconds) is 1000 divided by f. If you wanted to use a different frequency of pulses, just change the value f and the rest of the calculations will be taken care of automatically.

❹   When changing the PWM duty cycle, we need to set it to a value between 0 and 100, so the constant k is defined as 100 over the period, so that the constant can be used to scale an angle to a duty value.

❺   To convert the pulse length for zero angle to a corresponding value of duty between 0 and 100, the pulse length is multiplied by k.

❻   Similarly, the range of duty values is also calculated by multiplying the span of pulse lengths pulse_range by k.

❼ The GPIO pin is then set up and PWM starts running.

❽ The `set_angle` function converts the angle into a duty cycle value and then calls `ChangeDutyCycle` to set the new pulse length.

❾ The main loop is very similar to the Arduino version of this program, prompting for an angle and then setting it.

Experimenting with Raspberry Pi

Run the program using this command:

```
$ sudo python servo.py
```

Interacting with the program is much like controlling the servo from an Arduino using the Serial Monitor. Enter angles and watch the servo whizz around to the correct angle:

```
$ sudo python servo.py
Enter angle (0 to 180): 90
Enter angle (0 to 180): 0
Enter angle (0 to 180): 180
Enter angle (0 to 180): 90
Enter angle (0 to 180):  0
```

For the most part, the servo will respond just fine, but if you watch it for a while, you may notice the odd little jitter, especially if your Raspberry Pi is also doing a lot of other things. This is to be expected, because every so often the Raspberry Pi will generate a longer pulse than usual as its processor gets diverted to some other activity.

If the servo jitters too much, then an alternative to generating the servo pulses on the Raspberry Pi using software is to use a hardware module such as Adafruit's 16-channel PWM/Servo Driver board (product 815). This module uses just two pins to interface with the Raspberry Pi and allows control of up to 16 servos using a Python library also supplied by Adafruit.

Project: Pepe, the Dancing Raspberry Pi Puppet

Servomotors respond really quickly to commands to move from one position to another. In this project, they are used to pull the strings of a puppet and make it dance or move in any way you want (Figure 9-9).

Figure 9-9 *Pepe the Pi puppet*

The project is just controlled by editing a list of movements for the motors and program steps through the list of movements rather like playing a recording. In "Project: Pepe the Puppet Gets a Voice" on page 293, the project is expanded to give the puppet a voice, and in Chapter 16, it is given an Internet dimension, making the puppet dance in response to someone tweeting a certain hashtag.

Parts List

To make this project, you will need:

Part	Sources
Adafruit 16-Channel 12-bit PWM/Servo Driver	Adafruit: 815
4 x 9g Servo	eBay, Adafruit: 169
Female barrel jack to screw terminal adapter	Adafruit: 368
Female-to-female jumper wires	Adafruit: 266
Male-to-male jumper wires	Adafruit: 758
5V at 2A power supply	Adafruit: 276
4 x cocktail sticks (about 3 inches)	Supermarket
Small puppet (a string for each limb)	eBay
A4 or letter-sized sheet of mounting board	Craft shop
Hot glue gun or epoxy glue and a drill	Hardware store

Design

The project uses a Raspberry Pi, mainly so it can be adapted to communicate with the Internet in Chapter 16. However, you could also use an Arduino; if you do use an Arduino, you can lose the Adafruit controller board and connect the servos to the Arduino via a breadboard and male-to-male jumper wires. Lots of male-to-male jumper wires!

The project uses four small 9g servos, one for each arm and leg of the puppet. To make controlling the servos easy, an Adafruit 16-channel servo controller board is used. This has the added advantage that the servo lead sockets can just be plugged directly into the pins on the servo controller board.

Figure 9-10 shows the schematic diagram for the project.

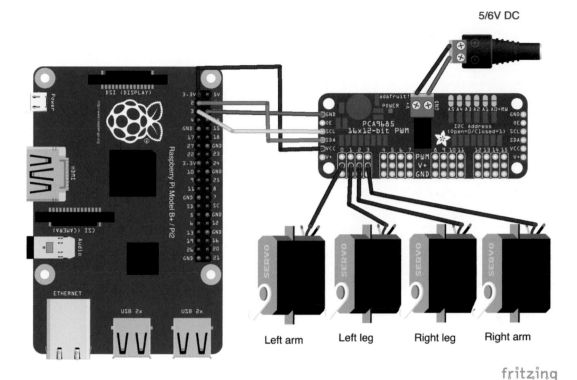

Figure 9-10 *Schematic diagram for the puppet project*

The board is designed to allow the servomotor sockets to push directly over the header pins (to prevent clutter, just the middle servo lead is shown in Figure 9-10).

Four servomotors jumping about all over the place at the same time means you will need a separate power supply for the motors so that motor noise does not interfere with the reliable operation of the Raspberry Pi.

The plastic arms supplied with the servomotors are not long enough for our purposes as you want a full range of movement for the puppet's arms and legs, so the plastic arms are lengthened using wooden cocktail sticks.

Construction

This project is as much about mechanical construction as electronics or software. Here are the steps needed to build your dancing puppet.

Step 1: Extend the servo arms

Servomotors come with a little bag of different types of plastic arm that will fit onto the servomotor. Pick one of the straighter arm designs and glue the cocktail stick to it as shown in Figure 9-11. A hot glue gun or epoxy glue will work best.

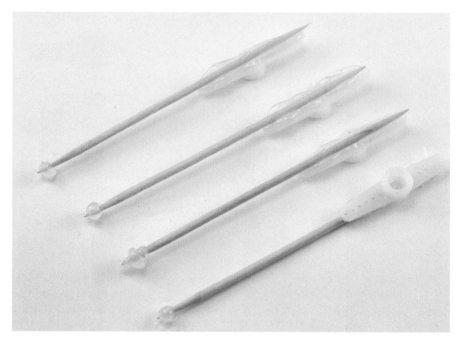

Figure 9-11 *Extending the servo arms*

Notice that I have also put a little blob of glue on the end of each stick to stop the puppet string from sliding off the end of the stick when it is tied to it.

Step 2: Make a chassis

The four servomotor arms need to be able to move freely up and down to control the movement of the limbs. To keep everything in the right place, I cut sections out of some mounting board (the kind that is used to mount photographs works well).

To work out where to make the cuts, print out the template found in *puppet.svg* in the directory *python/projects/puppet* and then stick it onto the board before cutting out the sections using a craft knife (Figure 9-12).

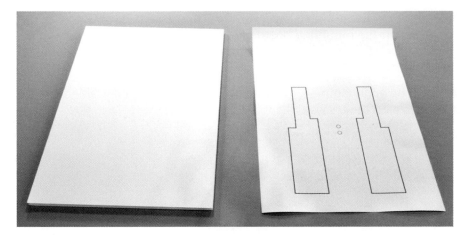

Figure 9-12 *Using a template for the chassis*

Stick the paper onto the board with "low-tack" glue so that you can peel it off again before you stick the motors to the chassis. As well as cutting out the two large areas in the board, you also need to drill two small holes that are needed for the string that goes to the puppet's head and supports the main weight of the puppet (Figure 9-13).

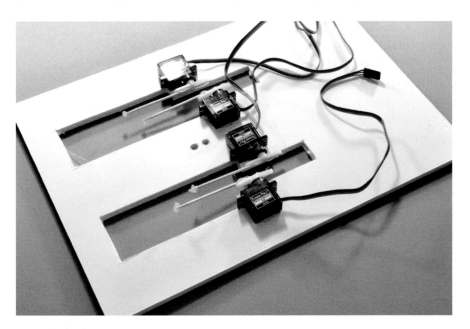

Figure 9-13 *Finishing the chassis*

Step 3: Glue the servos

Peel the paper template off the mounting board and then push the modified servo arms onto the servomotors. Don't fix them on too tight just yet, as they will need to come off again so that you can get the servos in the right position.

Line up the servos as shown in Figure 9-14 so that all the servo arms are able to move freely without getting in each other's way.

Figure 9-14 *Attaching the servos*

Before sticking the servos into place, you may want to remove the label on one side of the servo so that the servos stick better.

Step 4: Adapt the puppet

Figure 9-15 shows the puppet that I used.

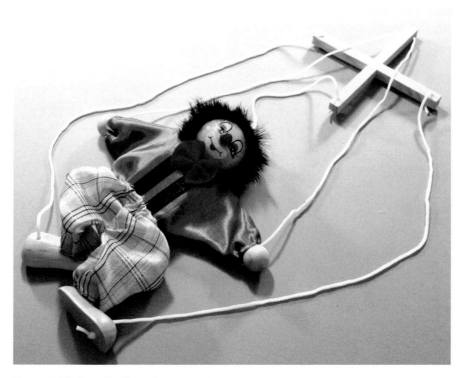

Figure 9-15 *Pepe the Puppet*

The puppet has one string from the top of his head that supports all of his weight. This is the string that will be threaded through the two holes in the chassis and tied underneath.

Each of the puppet's legs has a separate string and the arms are connected together with a single string that passes through one of the arms of the wooden cross. This last lead needs to be cut at the midpoint. Using a match or hot air gun on the end of the nylon-based string will melt it enough to stop the end fraying.

If you can just untie the knots on the end of the strings connected to the wooden cross, that's better than cutting them.

Before you tie the puppet to its new chassis, you first need to wire up the servos and run some software to get all the servo arms in the right position.

Step 5. Wire everything up

The Adafruit module is sold as a kit that is mostly assembled, but just needs the header pins soldered on. You can find full instructions for doing this at the Adafruit page for this product: *http://www.adafruit.com/products/815*.

I attached the header pins for the I2C interface to the Raspberry Pi on the opposite side of the servo controller board so that it will fit nicely on a breadboard when the project hardware is reused in Chapter 15 to give Pepe a voice.

If you want to reduce the amount of soldering you need to do, you can just solder on the header pins for the first four servos as this project only uses these four.

Wire everything up, referring back to Figure 9-10, and the project should look something like Figure 9-16.

Figure 9-16 *The completed wiring for the puppet project*

Make sure that the servo sockets are the right way around, with the orange control lead toward the top and the brown or black ground lead at the bottom.

At this point, you can now attach the 5V power supply and the seprate USB power supply for the Raspberry Pi.

Step 6: Run the test program

To get the servos in the right place, you need to use a separate program (*set_servos.py*) that allows you to set all the servo arms to a certain angle. This program and the main program for the puppet can be found in the folder */python/projects/puppet*.

The Adafruit servo module uses the Raspberry Pi's I2C interface that is not enabled by default.

To set up your Pi to use I2C, see the sidebar "Setting Up I2C on Your Raspberry Pi" on page 171.

Setting Up I2C on Your Raspberry Pi

Start by making sure that your package manager is up to date by using this command:

```
$ sudo apt-get update
```

Next run the raspi-config tool:

```
$ sudo raspi-config
```

From the menu that appears, choose "Advanced" and then "I2C." You will then be prompted with "Would you like the ARM I2C interface to be enabled?" Say "yes." You will then be asked "Would you like the I2C kernel module to be loaded by default?" This is an option that you would like to avail yourself of, so say "yes" again. Select "Finish" to exit raspi-config.

Run the following command from your home directory to install some useful I2C tools:

```
sudo apt-get install python-smbus i2c-tools
```

You can check that your Raspberry Pi is connected to the Adafruit servo board and ready to go by running this command:

```
$ sudo i2cdetect -y 1
```

You should see the following output (the numbers 40 and 70 in the table indicate that the board is connected):

```
$ sudo i2cdetect -y 1
     0  1  2  3  4  5  6  7  8  9  a  b  c  d  e  f
00:          -- -- -- -- -- -- -- -- -- -- -- --
10: -- -- -- -- -- -- -- -- -- -- -- -- -- -- -- --
20: -- -- -- -- -- -- -- -- -- -- -- -- -- -- -- --
30: -- -- -- -- -- -- -- -- -- -- -- -- -- -- -- --
40: 40 -- -- -- -- -- -- -- -- -- -- -- -- -- -- --
50: -- -- -- -- -- -- -- -- -- -- -- -- -- -- -- --
60: -- -- -- -- -- -- -- -- -- -- -- -- -- -- -- --
70: 70 -- -- -- -- -- --
```

Once you have set up I2C, remove the arms from the servos and run the program *set_servos.py*, and when prompted enter an angle of 90:

```
$ sudo python set_servos.py
Angle:90
Angle:
```

The servos should whir as they set their angle to 90. Now you can reattach the four servo arms so that they are as close to horizontal as the cogs on the servo shaft will allow.

Step 7: Attach the puppet

Now that you have all the servos in their 90° position, tie the head string to the holes in the middle of the chassis and then tie each of the puppet's limb strings to a servo arm so that the arms and feet are both half-raised.

At this point, you can run *set_servos.py* again and enter different angles to check that you have a full range of movement for the limbs.

Software

The software for this project is really just a starting point. It uses an array of servo positions into which you can put your own data to control the puppet's movement. The basic "dance" provided is certainly enthusiastic but could not be called elegant.

You might want to try out the program before taking a look at the code. You can find the program in the *dance.py* file inside the *puppet/* directory. Remember to run it using sudo:

```python
from Adafruit_PWM_Servo_Driver import PWM    # ❶
import time

pwm = PWM(0x40)

servoMin = 150  # Min pulse length out of 4095      # ❷
servoMax = 600  # Max pulse length out of 4095

dance = [           ❸
  #lh  lf  rf  rh
  [90, 90, 90, 90],
  [130, 30, 30, 130],
  [30, 130, 130, 30]
]

delay = 0.2     ❹

def map(value, from_low, from_high, to_low, to_high):  # ❺
  from_range = from_high - from_low
  to_range = to_high - to_low
  scale_factor = float(from_range) / float(to_range)
  return to_low + (value / scale_factor)


def set_angle(channel, angle):      ❻
  pulse = int(map(angle, 0, 180, servoMin, servoMax))
  pwm.setPWM(channel, 0, pulse)

def dance_step(step):       ❼
  set_angle(0, step[0])
  set_angle(1, step[1])
  set_angle(2, step[2])
  set_angle(3, step[3])


pwm.setPWMFreq(60)      ❽


while (True):       ❾
  for step in dance:
      dance_step(step)
      time.sleep(delay)
```

As well as the program itself (*dance.py*) the folder also contains some Adafruit files that are used by the program. The original source for these is also on GitHub (*https://github.com/ adafruit/Adafruit-Raspberry-Pi-Python-Code*).

❶ The Adafruit board is not just for servomotors; its outputs can also be used for PWM control of LEDs and other output devices. This is why the code imports a class called PWM.

❷ The two constants servoMin and servoMax are pulse lengths between 0 and 4095 where 4095 is 100% duty. This range of values should suit most servos and give an angle range of nearly 180°.

❸ The dance array contains the three steps for this particular dance. Add as many lines as you like here. Each line is made up of an array of four values that are angles for the left hand, left foot, right foot, and right hand, respectively. Because the arm and leg servos are mounted opposite ways around, an angle over 90 lifts an arm up, but moves a leg down (something to consider when planning your choreography).

❹ The delay variable sets the time delay between each step of the dance. The smaller the number, the faster the puppet will move.

❺ The map function is explained more in "The Arduino map Function" on page 253 in Chapter 12 and is used to expand the angle value to the right value of pulse length for use in the set_angle function.

❻ set_angle sets the position of the servo channel (0 to 3) in its first parameter to be the angle specified in its second parameter.

❼ The dance_step function takes the angles for the four limb servos and sets each servo to that angle.

❽ Setting the PWM frequency to 60 times a second gives a train of pulses every 17 milliseconds, which is about right for a servo.

❾ The main loop iterates through each step in the dance, setting the servos to the angles specified and then pausing for delay before moving on to the next step. When all the steps have been made, it starts again.

Using Pepe the Puppet

Try altering the contents of the dance array to add in your own movements for the puppet. You might like to try getting it to walk, wave, or stand on one leg. Setting the value of delay a bit higher will give you more time to see how the puppet is moving and to make adjustments to the dance array.

We'll work with Pepe again in Chapter 15 (where he is given a voice) and in Chapter 16 (where we make him dance in response to tweets).

Summary

Servomotors can be a lot of fun, and it's pretty easy to program them and connect mechanical things to them.

In the next chapter, you will learn about a completely different type of motor called the stepper motor.

Stepper Motors

10

If you have a standard printer (or, for that matter, a 3D printer), it probably contains a stepper motor or several. Figure 10-1 is responsible for feeding the plastic into the extruder of the 3D printer. Stepper motors are commonly used in printers because they move in a very precise way, one step at a time.

Figure 10-1 *A stepper motor in a 3D printer*

Quite different techniques are required to control stepper motors compared to DC motors. In this chapter, both unipolar and bipolar stepper motors are used, and the various driver chips that you will need to drive the motors are explored.

Stepper Motors

As the name suggests, stepper motors rotate in a series of short steps. This has the advantage that if the motor is free to turn, then by counting the steps you tell it to turn through, you know how much it has turned. This is one reason why stepper motors get used in printers, both of the 2D and 3D variety, to accurately position paper, a 3D printer bed, or print nozzle. Figure 10-2 shows three different types of stepper motor.

Figure 10-2 *A selection of stepper motors*

The tiny motor on the left is of the type used to move lens elements in a compact camera or smartphone camera. In the center is a 5V geared stepper motor. This combines both a stepper motor and a gearbox. On the right is a stepper motor typical of the sort you would find in a printer.

Bipolar Stepper Motors

Figure 10-3 shows how a stepper motor works. More specifically, it shows how a bipolar stepper motor works. In "Unipolar Stepper Motors" on page 190, you will learn about a different type of stepper motor.

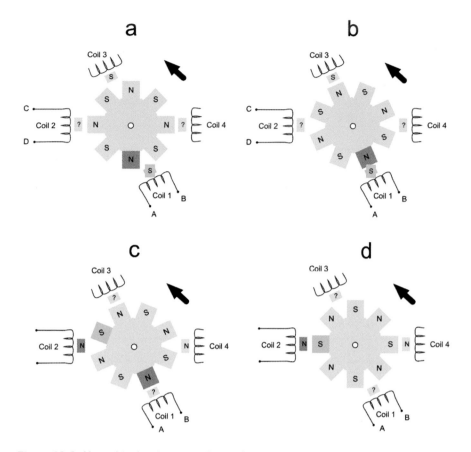

Figure 10-3 *How a bipolar stepper motor works*

Typically, there will be four coils, the opposite coils being connected together so that they work in unison. The coils are all on the stator (outside of the motor) and therefore there is no need for a commutator and brushes as there is in a DC motor.

In a stepper motor, the rotor is made in the form of a cog with alternating north and south poles marked by N and S. There will normally be a lot more cogs on the rotor than the number shown in Figure 10-3. Each of the coils can be energized to be magnetic north or south depending on the direction of the current through the coil (do you smell H-bridges in the air?). Coils 1 and 3 operate together, so that if Coil 1 is N so will Coil 3 be. The same applies to Coils 2 and 4.

Starting with Figure 10-3a, if Coil 1 (and therefore Coil 3) is energized to be S then opposites attract and like poles repel, so the rotor will rotate conter-clockwise until the S of Coil 1 is aligned with the nearest N cog, as shown in Figure 10-3b. To keep the clockwise turning, the next step is to energize Coils 2 and 4 to be N (Figure 10-3c), pulling the nearest S cog to Coil 2 level with the coil.

After all this effort, the motor has now advanced one step. To continue rotating counter-clockwise, Coil 1 now needs to be N.

Table 10-1 shows the sequence of coil activations to rotate the motor counter-clockwise.

Table 10-1 *Sequence for counter-clockwise rotation of a stepper motor*

Coil 1 and 3	Coil 2 and 4
S	-
-	N
N	-
-	S

A dash indicates that the coil is not having any effect on the rotation and does not need to be energized. In the cases where there is a dash, the polarity could be arranged to be the same as the cog that will be directly opposite it, to give an extra push to the motor, giving the revised table (Table 10-2).

Table 10-2 *Revised sequence of activations for rotation of a stepper motor*

Coil 1 and 3	Coil 2 and 4
S	N
N	N
N	S
S	S

You may be wondering what would have happened if we had started with Coil 2 rather than Coil 1. In that case, the other coil being energized will encourage the cog in the correct direction.

To reverse the direction of the rotor, all you need to do is energize the coils in the reverse of the table order in Table 10-2.

Experiment: Controlling a Bipolar Stepper Motor

There are two coils (well, actually two pairs, for a total of four coils altogether) to control and we need to reverse the direction of the current in both coils, so what we need is two H-bridges. This sounds like a job for the L293D.

In this experiment (shown in Figure 10-4), you will use an L293D on the breadboard to control a bipolar stepper motor using both Arduino and Raspberry Pi.

Figure 10-4 *Controlling a bipolar stepper motor*

Although this experiment uses a 12V stepper motor, it will still work with a 6V battery box to supply power to the motor. It will have less torque, but will turn just as well.

Identifying Stepper Motor Leads

If you are buying a new stepper motor, there should be an accompanying datasheet or even a label on the motor identifying each of the four pins. Most importantly, you need to know which leads go together as a pair, connected to the same coil.

A trick for working this out is to pick any two wires and hold them together between thumb and forefinger while you twist the shaft of the motor. If you can feel resistance in turning the shaft, those two wires are a pair.

Parts List

Whether you are using a Raspberry Pi or Arduino (or both), you are going to need the following parts to carry out this experiment:

Name	Part	Sources
IC1	L293D H-bridge IC	Adafruit: 807 Mouser: 511-L293D
C1	100nF capacitor	Adafruit: 753 Mouser: 810-FK16X7R2A224K
C2	100μF 16V capacitor	Adafruit: 2193 Sparkfun: COM-00096 Mouser: 647-UST1C101MDD
M1	Bipolar stepper motor 12V	Adafruit: 324
	6V (4 x AA) battery box	Adafruit: 830
	400-point solderless breadboard	Adafruit: 64
	Male-to-male jumper wires	Adafruit: 758
	Female-to-male jumper wires (Raspberry Pi only)	Adafruit: 826

If you are planning to try this experiment with a Raspberry Pi, you will need female-to-male jumper wires to connect the Raspberry Pi GPIO pins to the breadboard.

Design

Figure 10-5 shows the schematic diagram for this experiment.

In this case, we are not going to use PWM and so the two Enable pins of the L293D are connected to 5V to leave both H-bridges permanently enabled. The stepper motor is controlled by the four connections In 1, In 2, In 3, and In 4 that will be driven by digital outputs on the Arduino or Raspberry Pi.

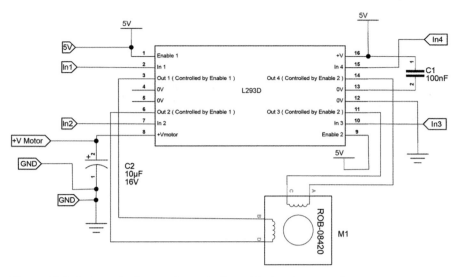

Figure 10-5 *Schematic diagram for controlling a bipolar stepper motor*

Arduino

The Arduino version of this experiment will use the Serial Monitor to send commands to the Arduino that will then control the motor. The three commands are represented by a single letter followed by a number. For example:

- f100 moves the motor forward by 100 steps
- r100 moves the motor 100 steps in the reverse direction
- p10 sets the period of the delay between step pulses to be 10 milliseconds

Figure 10-6 shows the Arduino version of the experiment.

Arduino Connections

The Arduino version of this experiment uses the following pins connected to the L293D:

L293D pin name	L293 pin number	Arduino pin
In1	2	10
In2	7	9
In3	10	11
In4	15	8

Figure 10-7 shows the breadboard layout for Arduino.

Figure 10-6 *Arduino stepper motor control*

Make sure that the IC is positioned correctly (the notch should be at the top), and that C2 the 100µF capacitor has its positive lead connected to pin 8 of the L293D. The positive lead of an electrolytic capacitor like C2 is normally longer than the negative lead. The negative lead of C2 may also be marked on the body of the capacitor with a minus sign or diamond symbol.

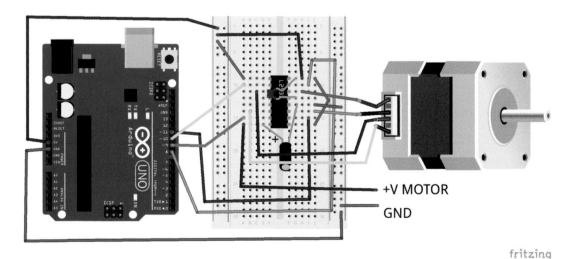

+V MOTOR

GND

fritzing

Figure 10-7 *Arduino breadboard layout for a bipolar stepper motor*

Arduino Software (the Hard Way)

There are two versions of the Arduino software for this experiment. The first does things the hard way, by setting the control pins for the L293D in the sequence described in "Bipolar Stepper Motors" on page 176. This is a useful exercise for understanding exactly how the stepper motor works.

The second example sketch uses the Arduino Stepper library to do all the work and is therefore a lot shorter.

The first Arduino sketch can be found in *arduino/experiments/bi_stepper_no_lib*:

```
const int in1Pin = 10;      ❶
const int in2Pin = 9;
const int in3Pin = 11;
const int in4Pin = 8;

int period = 20;       ❷

void setup() {       ❸
  pinMode(in1Pin, OUTPUT);
  pinMode(in2Pin, OUTPUT);
  pinMode(in3Pin, OUTPUT);
  pinMode(in4Pin, OUTPUT);
  Serial.begin(9600);
  Serial.println("Command letter followed by number");
  Serial.println("p20 - set the inter-step period to 20ms (control speed)");
  Serial.println("f100 - forward 100 steps");
  Serial.println("r100 - reverse 100 steps");
}

void loop() {       ❹
  if (Serial.available()) {
    char command = Serial.read();
    int param = Serial.parseInt();
    if (command == 'p') {       ❺
      period = param;
    }
    else if (command == 'f') {       ❻
      stepForward(param, period);
    }
    else if (command == 'r') {
      stepReverse(param, period);
    }
  }
  setCoils(0, 0, 0, 0); // power down
}

void stepForward(int steps, int period) {       ❼
  for (int i = 0; i < steps; i++) {
    singleStepForward(period);
  }
}
```

```
void singleStepForward(int period) {        ❽
  setCoils(1, 0, 0, 1);
  delay(period);
  setCoils(1, 0, 1, 0);
  delay(period);
  setCoils(0, 1, 1, 0);
  delay(period);
  setCoils(0, 1, 0, 1);
  delay(period);
}

void stepReverse(int steps, int period) {
  for (int i = 0; i < steps; i++) {200
    singleStepReverse(period);
  }
}

void singleStepReverse(int period) {        ❾
  setCoils(0, 1, 0, 1);
  delay(period);
  setCoils(0, 1, 1, 0);
  delay(period);
  setCoils(1, 0, 1, 0);
  delay(period);
  setCoils(1, 0, 0, 1);
  delay(period);
}

void setCoils(int in1, int in2, int in3, int in4) {     // ❿
  digitalWrite(in1Pin, in1);
  digitalWrite(in2Pin, in2);
  digitalWrite(in3Pin, in3);
  digitalWrite(in4Pin, in4);
}
```

❶ The code starts by defining the pins that will be used to control the motor. You can, of course, change these pins as long as you also modify your breadboard to match.

❷ The variable period is the delay between each coil activation in a step of the motors' rotation. This is initially set to 20 milliseconds. You can change this value over the Serial Monitor to speed up or slow down the motor.

❸ The setup() function sets the control pins to be digital outputs, starts serial communication with the Serial Monitor, and then prints out a message explaining the format of the commands that you can send.

❹ The loop() function waits for commands to arrive over serial and then processes them. If a command has been received (indicated by Serial.available), then loop() first reads the command letter into the variable command and then reads the number parameter that follows it.

Next, a series of if statements take the appropriate action depending on the command.

❺ If the command is 'p' then the period variable is set to the number supplied as a parameter.

❻ If, however, the command is 'f' or 'r' then stepForward or stepReverse is called as appropriate accompanied by the number of steps to take and the period between each polarity change in the step as parameters.

❼ stepForward and stepReverse are very similar and repeatedly call singleStepForward or singleStepReverse for the number of times specified in steps.

❽ Next, we come to the function singleStepForward that holds the pattern of polarities needed for the four phases to advance the motor by one step. You can see the pattern in the parameters to setCoils.

❾ singleStepReverse is the same as singleStepForward but reverses the sequence. Compare the steps with Table 10-2.

❿ Finally, the setCoils function sets the control pins according to the pattern supplied as its parameters.

Arduino Software (the Easy Way)

The Arduino IDE includes the Stepper library, which works just fine and will reduce the size of the sketch considerably. You can find this sketch in the *arduino/experiments/ex_07_bi_stepper_lib/* directory (you'll find this in the place where you downloaded the book's code—see "Downloading the Software" on page 45):

```
#include <Stepper.h>      ❶

const int in1Pin = 10;
const int in2Pin = 9;
const int in3Pin = 8;
const int in4Pin = 11;

Stepper motor(200, in1Pin, in2Pin, in3Pin, in4Pin);     // ❷

void setup() {        ❸
  pinMode(in1Pin, OUTPUT);
  pinMode(in2Pin, OUTPUT);
  pinMode(in3Pin, OUTPUT);
  pinMode(in4Pin, OUTPUT);
  while (!Serial);
  Serial.begin(9600);
  Serial.println("Command letter followed by number");
  Serial.println("p20 - set the motor speed to 20");
  Serial.println("f100 - forward 100 steps");
  Serial.println("r100 - reverse 100 steps");
```

```
    motor.setSpeed(20);          ❹
  }

  void loop() {                 ❺
      if (Serial.available()) {
      char command = Serial.read();
      int param = Serial.parseInt();
      if (command == 'p') {
        motor.setSpeed(param);
      }
      else if (command == 'f') {
        motor.step(param);
      }
      else if (command == 'r') {
        motor.step(-param);
      }
    }
  }
```

❶ The first line of the sketch imports the Stepper library, which is included as part of the Arduino IDE and does not need to be specially installed.

❷ To use the library, a variable motor is defined of type Stepper. The parameters supplied to motor set up the motor so that it can be used. The first parameter is the number of steps the motor takes per revolution. The Adafruit stepper motor that I used is specified as 200 steps, which is a common number of steps. Actually this really means 200 phase changes per revolution, which is 50 actual shaft positions per revolution. The other four parameters are the coil pins to use.

❸ The setup() function is almost the same as the long version of the sketch. The message is slightly different, because in this case, the 'p' command will set the motor's speed in rpm (revolutions per minute) rather than the raw delay time. However, both have the effect of changing the motor's speed.

❹ The default speed is also set to 20rpm in setup() using motor.setSpeed.

❺ The loop() function is also very similar to the long version of the sketch; however, in this case, the number of steps and direction for the motor are specified as a positive number for forward and a negative number for reverse.

The number of quarter-steps specified in the 'f' or 'r' parameter is in this case the number of phase changes, so for one complete revolution of the Adafruit motor you will need to enter a value of 200.

Arduino Experimentation

Try out both versions of the stepper code by all means, but here I will assume that you have *ex_05_bi_stepper_lib* uploaded to your Arduino. Open the Serial Monitor and you should be greeted with the message shown in Figure 10-8.

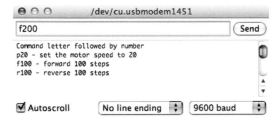

Figure 10-8 *Controlling the stepper motor in the Serial Monitor*

Enter the command f200 and click Send. The motor should turn through one complete revolution. If it doesn't turn, but rather jitters or hums, it probably means that you positioned one of the coil leads incorrectly. So, swap over the stepper leads that go to pins 3 and 6 of the L293D and try the command again.

Try r200 and the motor should turn through one revolution in the opposite direction.

Next, try experimenting with the speed using the command p followed by the speed in rpm. For example, p5 followed by f200. Again the motor will turn one revolution, but very slowly. You can increase the speed by specifying a higher value, say p100, but you will find that there is a maximum speed somewhere around 130rpm before some steps get lost and the motor does not make one complete revolution.

Raspberry Pi

Controlling a stepper motor using Python or a Raspberry Pi is very similar to the non-library Arduino approach. Controlling the L293D will require four control outputs from the Raspberry Pi.

The Python program will prompt you to enter the inter-phase delay, direction, and number of steps to take.

Raspberry Pi Connections

You can keep the breadboard layout as the Arduino, but the header pins for connecting to the Raspberry Pi need to be female-to-male rather than male-to-male. Figure 10-9 shows the breadboard layout to connect to a Raspberry Pi.

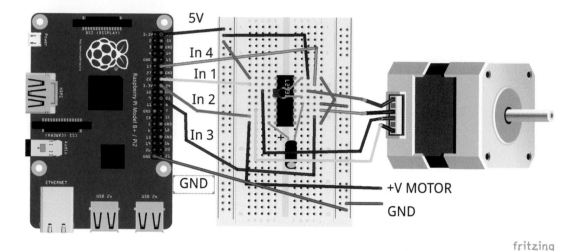

5V

In 4

In 1

In 2

In 3

GND

+V MOTOR

GND

fritzing

Figure 10-9 *Raspberry Pi breadboard layout for a stepper motor*

Raspberry Pi Software

The Raspberry Pi version of the stepping code is almost a word-for-word translation of the non-library Arduino code. This program is called *bi_stepper.py* and is located in the *python/experiments* directory (you'll find this in the place where you downloaded the book's code):

```python
import RPi.GPIO as GPIO
import time

GPIO.setmode(GPIO.BCM)

in_1_pin = 23          ❶
in_2_pin = 24
in_3_pin = 25
in_4_pin = 18

GPIO.setup(in_1_pin, GPIO.OUT)
GPIO.setup(in_2_pin, GPIO.OUT)
GPIO.setup(in_3_pin, GPIO.OUT)
GPIO.setup(in_4_pin, GPIO.OUT)

period = 0.02

def step_forward(steps, period):    ❷
  for i in range(0, steps):
    set_coils(1, 0, 0, 1)
    time.sleep(period)
    set_coils(1, 0, 1, 0)
    time.sleep(period)
    set_coils(0, 1, 1, 0)
    time.sleep(period)
    set_coils(0, 1, 0, 1)
    time.sleep(period)
```

```
def step_reverse(steps, period):
  for i in range(0, steps):
    set_coils(0, 1, 0, 1)
    time.sleep(period)
    set_coils(0, 1, 1, 0)
    time.sleep(period)
    set_coils(1, 0, 1, 0)
    time.sleep(period)
    set_coils(1, 0, 0, 1)
    time.sleep(period)


def set_coils(in1, in2, in3, in4):          ❸
  GPIO.output(in_1_pin, in1)
  GPIO.output(in_2_pin, in2)
  GPIO.output(in_3_pin, in3)
  GPIO.output(in_4_pin, in4)


try:
    print('Command letter followed by number');
    print('p20 - set the inter-step period to 20ms (control speed)');
    print('f100 - forward 100 steps');
    print('r100 - reverse 100 steps');

    while True:          ❹
        command = raw_input('Enter command: ')
        parameter_str = command[1:] # from char 1 to end
        parameter = int(parameter_str)          ❺
        if command[0] == 'p':          ❻
            period = parameter / 1000.0
        elif command[0] == 'f':
            step_forward(parameter, period)
        elif command[0] == 'r':
            step_reverse(parameter, period)

finally:
    print('Cleaning up')
    GPIO.cleanup()
```

❶   The program defines constants for the four control pins and sets them to be outputs.

❷   The step_forward and step_reverse functions are just like their equivalents in the Arduino. The four coil activations take place in the appropriate order using the set_coils function. This is repeated steps times with a delay of period between each change in coil activation.

❸   set_coils sets the digital output control pins according to its four parameters.

❹   The main loop of the program reads the command string using raw_input. The parameter that follows the letter is first cut off the command using the syntax [1:], which for a Python string means the string from position 1 (the second character) to the end of the string.

❺ This string parameter is then converted into a number using the `int` built-in function.

❻ A series of three `if` statements then carry out the appropriate action for the command, either changing the value of the `period` variable or sending the motor in one direction or the other.

Raspberry Pi Experimentation

Run the Python program using the command `sudo python ex_07_bi_stepper.py`. You will be prompted with the possible commands that you can enter. These are just the same as for the Arduino version.

Test out running the motor in both directions and find the smallest value of delay that you can achieve before the motor starts missing steps:

```
$ sudo python ex_05_bi_stepper.py
Command letter followed by number
p20 - set the inter-step period to 20ms (control speed)
f100 - forward 100 steps
r100 - reverse 100 steps
Enter command: p5
Enter command: f50
Enter command: p10
Enter command: r100
Enter command:
```

Unipolar Stepper Motors

Unipolar stepper motors work in a very similar manner to bipolar motors, but do not need H-bridges to control them. The accomplish this trick by a more complicated coil arrangement. Figure 10-10 shows how a unipolar stepper motor works.

A unipolar stepper motor will have five leads rather than the four of a bipolar motor. Four of the leads are just the same as those of a bipolar motor; they are just the ends of the coils, A, B, C, and D. In fact, there will be another pair of coils, not shown in Figure 10-10, just like a bipolar motor. If you want to, you can use the leads A, B, C, and D with an H-bridge just as if it is a bipolar stepper and it will work just fine.

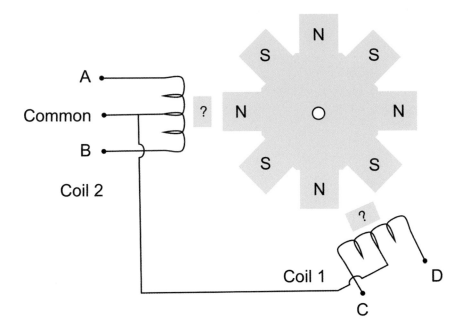

Figure 10-10 *A unipolar stepper motor*

The fifth lead connects half way along each of the coils. If you imagine that this lead was connected to ground, then you can polarize the coil to be North by supplying power through A or South by supplying power to B. No need for an H-bridge.

Darlington Arrays

Although you do not need an H-bridge to use a unipolar stepper motor, the motor coils still take too much current to be driven directly from the pins of an Arduino or Raspberry Pi.

The obvious way of upping the current is to use transistors, just as you did back in "Experiment: Controlling a Motor" on page 53. You will of course need four transistors, one for each of the coil connections A, B, C, and D. You will also need current limiting resistors connected to the bases of these transistors as well as protection diodes across each coil. Figure 10-11 shows the schematic that you might use to connect to each coil of a unipolar stepper. For a full system, you would need to repeat this four times.

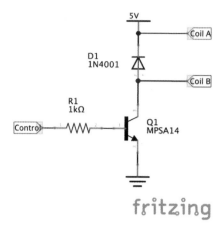

Figure 10-11 *Driving a unipolar stepper with separate transistors*

You can make this just fine on the breadboard, but there will be quite a lot of wires to connect up.

A neater approach is to use a driver chip such as the ULN2803. These low-cost ICs have eight Darlington transistors in one package, and also include diode protection and base resistors, so the only thing you need to add is a reservoir capacitor.

The ULN2803 (*http://www.adafruit.com/datasheets/ULN2803A.pdf*) can provide 500mA per channel and a maximum voltage of 50V.

In "Experiment: Controlling a Unipolar Stepper Motor" on page 192, you will use one of these devices to control a unipolar stepper motor from Arduino and Raspberry Pi.

Experiment: Controlling a Unipolar Stepper Motor

Figure 10-12 shows an Arduino controlling a unipolar stepper motor. The motor used is another widely available stepper motor type. It includes an integrated 1:16 reduction gearbox. So, although the motor itself only has 32 steps, the gearbox means that it actually takes 513 phase changes for one revolution, equating to 128 steps per revolution.

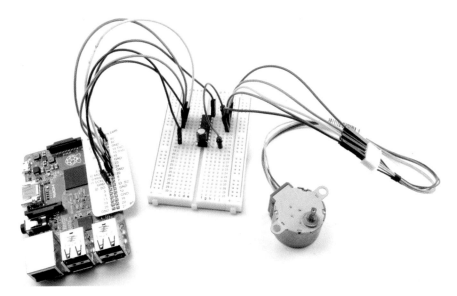

Figure 10-12 *Controlling a unipolar stepper motor with a Raspberry Pi*

Hardware

Figure 10-13 shows the circuit of Figure 10-11 reimplemented using a ULN2003. Note that there are still four unused Darlington transistors that could be used to control a second stepper motor. For higher-current stepper motors, you can even double-up the outputs by connecting two inputs together with the corresponding two outputs.

The motor used here is 5V and also low current enough to draw its power from the Raspberry Pi or Arduino. It draws about 150mA with two coils activated. If you have a hungrier motor, or one that operates at a higher voltage, you will need to use a separate power supply, as you did with the bipolar stepper motor.

Finding which leads are connected to what inside the motor can be done in the same way as for bipolar motors (see "Identifying Stepper Motor Leads" on page 179) with the added complication that there is a common lead. So, start by identifying the common lead, which will give resistance to turning when connected to any of the other leads. Then use the same method as for a bipolar motor on the remaining leads.

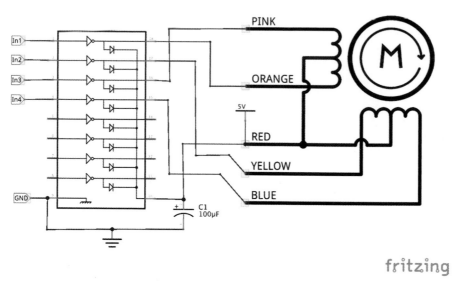

Figure 10-13 *Schematic for unipolar motor control (lead colors may vary)*

Parts List

Whether you are using a Raspberry Pi or Arduino (or both), you are going to need the following parts to carry out this experiment:

Name	Part	Sources
IC1	ULN2803	Adafruit: 970 Mouser: 511-ULN2803A
C2	100µF 16V capacitor	Adafruit: 2193 Sparkfun: COM-00096 Mouser: 647-UST1C101MDD
M1	Unipolar Stepper Motor 5V	Adafruit: 858
	6V (4 x AA) battery box	Adafruit: 830
	400-point solderless breadboard	Adafruit: 64
	Male-to-male jumper wires	Adafruit: 758
	Female-to-male jumper wires (Raspberry Pi only)	Adafruit: 826

If you are planning to try this experiment with a Raspberry Pi, you will need female-to-male jumper wires to connect the Raspberry Pi GPIO pins to the breadboard.

Arduino Connections

Figure 10-14 shows how an Arduino Uno can be connected up to a ULN2803.

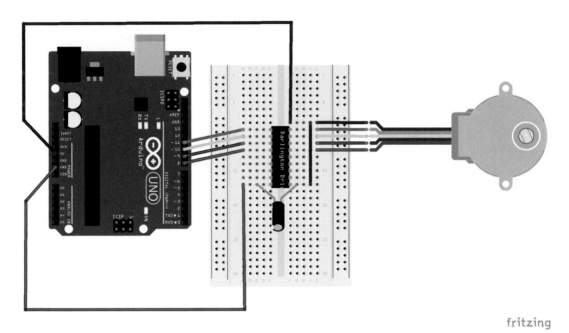

fritzing

Figure 10-14 *Connecting an Arduino to a ULN2803*

You need to make sure that the capacitor is positioned properly, with the longer positive lead to the right. The chip should be positioned with the notch toward the top of the board.

The layout is actually really neat compared with that of the bipolar layout.

Raspberry Pi Connections

Figure 10-15 shows the Raspberry Pi layout and connections.

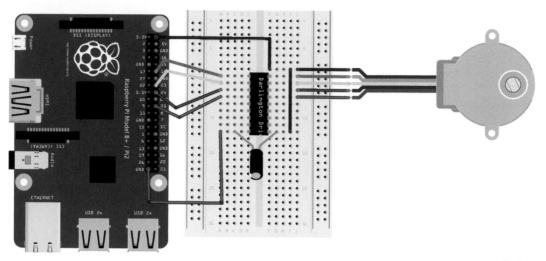

fritzing

Figure 10-15 *Connecting a Raspberry Pi to a ULN2803*

Software

Apart from the fact that some of the pins are swapped over, the software for both Arduino and Raspberry Pi is exactly the same as for "Experiment: Controlling a Bipolar Stepper Motor" on page 178. The programs with the modified pin connections are:

- For the Arduino: uni_stepper_lib.ino
- For the Raspberry Pi: uni_stepper.py

Microstepping

You probably noticed that the stepper motor's rotation was not very smooth, even when rotating fairly fast. For most applications, this isn't really a problem, but sometimes it's nice to have the movement be a bit smoother.

Microstepping is a technique that allows a much smoother motion of the motor. Rather than simply turning the coils on and off, PWM is used to energize the coils more smoothly. This creates a smoother rotation of the motor.

While possible using software, it is generally easier to use hardware specifically designed to microstep stepper motors such as the EasyDriver designed by Brian Schmalz. This board is based on the A3967 IC.

You can find a great tutorial on using this board with an Arduino on Sparkfun's project page for this board (product ROB-12779), so I will not repeat that information here. Instead, I offer a Raspberry Pi alternative in the form of "Experiment: Microstepping on Raspberry Pi" on page 197.

Experiment: Microstepping on Raspberry Pi

In this experiment, you will use an EasyDriver stepper motor board to microstep the 12V bipolar stepper motor that you used in "Experiment: Controlling a Bipolar Stepper Motor" on page 178 with a Raspberry Pi. You could also use a unipolar motor, but don't connect the common lead and use the unipolar motor as if it were bipolar.

The EasyDriver stepper motor board is supplied with header pins that allow it to be plugged directly onto breadboard.

Figure 10-16 shows the finished build for the experiment.

Figure 10-16 *Microstepping with a Raspberry Pi*

Parts List

You need the following parts to carry out this experiment:

Part	Sources
EasyDriver stepper motor board	Sparkfun: ROB-12779
Bipolar stepper motor 12V	Adafruit: 324
6V (4 x AA) battery box	Adafruit: 830
400-point solderless breadboard	Adafruit: 64
Female-to-male jumper wires	Adafruit: 826

Most 12V stepper motors will work OK at 6V, but feel free to substitute the battery pack for a 12V power supply.

Raspberry Pi Connections

Figure 10-17 shows the breadboard layout and connections to the Raspberry Pi.

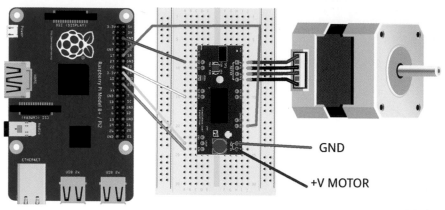

GND

+V MOTOR

fritzing

Figure 10-17 *Breadboard layout and connections for microstepping with a Raspberry Pi*

The EasyDriver board has the capacitors on board that you would normally need to add to your circuit. It also has a voltage regulator chip that provides the power supply to its A3967 IC, so you do not even need to supply the board with power from the Pi. All that is needed is a common ground and four control signals.

Software

As its name correctly suggests, the EasyDriver stepper motor is extremely easy to control using software. The main process of telling the motor to step is accomplished by just two control pins. When the step pin receives a HIGH pulse, the motor takes one step in the direction set by the direction pin.

Two other pins, ms1 and ms2, control the degree of microstepping from none to 1/8th.

The code for the following program is in the *microstepping.py* file (which you'll find in the place where you downloaded the book's code):

```
import RPi.GPIO as GPIO
import time

GPIO.setmode(GPIO.BCM)

step_pin = 24
dir_pin = 25
ms1_pin = 23
ms2_pin = 18
```

```
GPIO.setup(step_pin, GPIO.OUT)
GPIO.setup(dir_pin, GPIO.OUT)
GPIO.setup(ms1_pin, GPIO.OUT)
GPIO.setup(ms2_pin, GPIO.OUT)

period = 0.02

def step(steps, direction, period):      ❶
  GPIO.output(dir_pin, direction)
  for i in range(0, steps):
    GPIO.output(step_pin, True)
    time.sleep(0.000002)
    GPIO.output(step_pin, False)
    time.sleep(period)

def step_mode(mode):                      ❷
    GPIO.output(ms1_pin, mode & 1)        ❸
    GPIO.output(ms2_pin, mode & 2)

try:
    print('Command letter followed by number');
    print('p20 - set the inter-step period to 20ms (control speed)');
    print('m - set stepping mode (0-none 1-half, 2-quater, 3-eighth)');
    print('f100 - forward 100 steps');
    print('r100 - reverse 100 steps');

    while True:                           ❹
        command = raw_input('Enter command: ')
        parameter_str = command[1:] # from char 1 to end
        parameter = int(parameter_str)
        if command[0] == 'p':
            period = parameter / 1000.0
        elif command[0] == 'm':
            step_mode(parameter)
        elif command[0] == 'f':
            step(parameter, True, period)
        elif command[0] == 'r':
            step(parameter, False, period)

finally:
    print('Cleaning up')
    GPIO.cleanup()
```

By now, you should be familiar with the GPIO pin setup and initialization code that starts most of the Python programs in this book. If not, refer back to some of the earlier chapters.

❶   The function step contains the actual code for moving the motor on by a number of steps (or microsteps). It takes parameters of the number of steps to take, the direction (0 to 1), and the period of the delay to leave between each step. The body of the function first of all sets the direction pin (dir_pin) of the EasyDriver board and then generates the required number of pulses on the step_pin. The duration of each step_pin pulse is just 2 microseconds. The datasheet for the A3967 says that the pulse must be at least 1 microsecond.

❷ The step_mode function sets the stepping mode pins according to the value of mode that should be between 0 and 3.

❸ The code inside step_mode separates the two bits of the mode number and sets ms1_pin and ms2_pin using the logical and operator (&). Table 10-3 shows the relationship between the mode number, the microstepping pins, and the behavior of the motor.

Table 10-3 *Microstepping control pins*

Value of mode	ms1_pin	ms2_pin	Microstepping mode
0	LOW	LOW	None
1	HIGH	LOW	Half stepping
2	LOW	HIGH	Quarter stepping
3	HIGH	HIGH	One eighth stepping

❹ The main loop() function is very similar to that of the Python program for "Experiment: Controlling a Bipolar Stepper Motor" on page 178 except that an extra command of m is added to allow the microstepping mode to be set.

Experimenting

Run the program *microstepping.py* and then the program will prompt you to enter a command:

```
$ sudo python microstepping.py
Command letter followed by number
p20 - set the inter-step period to 20ms (control speed)
m - set stepping mode (0-none 1-half, 2-quater, 3-eighth)
f100 - forward 100 steps
r100 - reverse 100 steps
Enter command: m0
Enter command: p8
Enter command: f800
```

Enter the commands m0, p8, and then f800. This should result in the motor rotating through four revolutions (for a 200-step motor). Make a mental note of the smoothness (or otherwise) of the rotation. Next, enter the following commands: m3, p1, f6400.

The motor should still rotate four times and at the same speed but noticeably smoother.

Note that to keep the same speed, the period between steps was reduced by a factor of 8 and the number of steps (or in this case, microsteps) increased by a factor of 8.

Brushless DC Motors

Brushless DC (BLDC) motors (Figure 10-18) are the ferocious little high-powered motors that you find in quadcopters and remote control airplanes. Weight for weight, they can create far more torque than a regular DC motor. They are included in this chapter, because although technically a DC motor, they actually have more in common with stepper motors.

Figure 10-18 *A BLDC motor*

Unlike "brushed" DC motors (see Chapter 7), BLDC motors do not have a mechanical commutator that reverses the direction of the current as the motor rotates. Instead, they are more similar in design to a stepper motor, except that instead of having two coils (each in pairs) they have three, repeated many times. They can therefore be classified as three-phase motors, and driving these motors involves more than just an H-bridge. In fact, each of the three connections to a coil must be capable of sourcing current, sinking current, and also being in a third "high impedance" state where the coil connection is effectively disconnected. To drive the motor effectively, the motor driver circuit measures the voltage at the "disconnected" coil for a particular coil to adjust the time for switching to the next phase.

Sadly, there does not seem to be a readily available breadboard-able IC that can take on the role of driving such a motor, so the best bet is to look for a ready-made driver board, of which various models can be found on eBay.

If you want the power and size of a BLDC motor without the fuss of a driver board, you can also buy motors that have a built-in controller and just two wires exiting the motor housing, so that you can use it just like it is a DC motor. In fact, the velocity pump pictured back in Figure 7-13 uses just such a motor.

Summary

In this chapter, we have explored the topic of stepper motors and learned how they can be used with both Arduino and Raspberry Pi. In the next chapter, you will learn about how to control heating and cooling.

Heating and Cooling 11

This chapter explores devices for generating heat and also cooling. This includes resistive heater elements and Peltier elements.

This chapter focuses on how heaters and coolers work. In Chapter 12 we will use the heating and cooling devices described here with Arduino and Raspberry Pi for precise thermostatic control.

Resistive Heaters

Back in Chapter 5, we touched on the fact that in the process of restricting the flow of current, resistors generate heat. As you might recall, the amount of heat generated in watts is the current in amps flowing through the resistor multiplied by the voltage drop in volts across the resistor.

Experiment: Resistor Heating

You should now be familiar with simply turning power on and off to start a motor. Exactly the same principles apply when switching the power to a resistor. Any of the previous experiments that have controlled the power to a motor will work equally well with a resistor. Even using PWM to control the power will work just as well with a resistor heater. So, to carry out this experiment, you can dispense with your Arduino or Raspberry Pi and just make do with a battery pack and a resistor.

Note that this project uses a resistor as a heating element and a battery as a power source. Neither of these choices are very practical unless you are heating something very small. They are, however, very convenient for experimentation.

Parts List
You will need the following parts to try out this experiment:

Part	Sources
100Ω resistor	Mouser: 291-100-RC
Thermometer	Hardware store
400-point solderless breadboard	Adafruit: 64
6V (4 x AA) battery box	Adafruit: 830

Construction

In this project, you are really just connecting the resistor across the battery terminals and using the breadboard to hold the resistor in place, as shown in Figure 11-1.

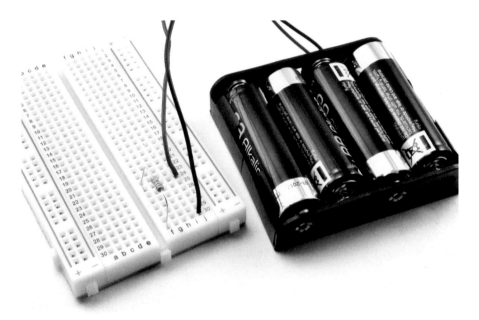

Figure 11-1 *Resistive heating experiment*

Don't attach the battery just yet, because as soon as you do, the resistor is going to get hot.

Experimenting

Connect the battery box, and use your thermometer to measure how quickly the temperature rises on the resistor. The resistor temperature will probably get up to about 170° F (75° C), so watch out—that's hot enough to burn your finger!

The resistor is 100Ω and the voltage 6V, from which we can calculate the current as I = V / R = 6 / 100 = 0.06A. The power is the voltage multiplied by the current, which is 6 x 0.06 = 0.36W, or 360mW.

That's a little more than the resistor's maximum power rating of 250mW (1/4W), so the resistor may fail after a while.

Project: Arduino Random Balloon Popper

This is not a project for the balloon-phobic. It uses an Arduino to control a resistor as a heating element that is taped to a balloon and after a random period of time, the resistor is powered up and pops the balloon (Figure 11-2).

Figure 11-2 *A random balloon popper*

This is a fairly destructive project, as to get the resistor hot enough to pop a balloon also means making it hot enough to damage itself—you are quite likely to see some smoke.

Danger of Burns
This project uses a resistor to pop a balloon, and that resistor gets hot enough to burn you fingers. You should not touch the resistor when it is turned on, and after you've turned the resistor off, you should wait at least 30 seconds before handling it. This project also poses a fire risk, so be sure to have a fire extinguisher handy just in case. In addition, the resistor might fly off with the explosion of the balloon, so make sure to be alert and wear safety goggles to protect your eyes.

Parts List

You are going to need the following parts to build this project:

Name	Part	Sources
R1, R3	470Ω resistor	Mouser: 291-470-RC
R2	10Ω 1/4 W resistor	Mouser: 291-10-RC
Q1	MPSA14 Darlington transistor	Mouser: 833-MPSA14-AP
LED1	Red LED	Adafruit: 297 Sparkfun: COM-09590
	400-point solderless breadboard	Adafruit: 64
	Male-to-male jumper wires	Adafruit: 758

You should get several 10Ω resistors, as these are likely to literally go up in smoke.

Hardware

By the time the Darlington transistor has taken its share, there will be about 3.5V across the resistor when the transistor turns on. This means a current of 3.5V / 10Ω = 350mA. This should not overload any USB device powering the project.

The heat power coming from the resistor will be I x V = 350mA x 3.5V = 1.225W. This is a lot more than the 1/4W that the resistor is rated at, but the resistor should survive long enough to burst the balloon.

Figure 11-3 shows the breadboard layout for the project.

The resistor should be firmly attached to the jumper wire pins (Figure 11-4), so that it does not come adrift when the balloon pops.

Figure 11-3 *The breadboard layout for the balloon popper*

Figure 11-4 *Making a resistor lead*

Software

There is no start button for this project apart from the Arduino's Reset button. The Arduino sketch for this project starts as soon as it's plugged in, so you should keep one of the jumper wires to the resistor disconnected until you are ready to do some popping.

Go to */arduino/projects/pr_balloon_popper* (which you'll find in the place where you downloaded the book's code):

```
const int popPin = 11;

const int minDelay = 3;  // Seconds ❶
const int maxDelay = 5;  // Seconds
const int onTime = 3;     // Seconds ❷

void setup() {
  pinMode(popPin, OUTPUT);
  randomSeed(analogRead(0));      ❸
  long pause = random(minDelay, maxDelay+1);   ❹
  delay(pause * 1000);          ❺
  digitalWrite(popPin, HIGH);      ❻
  delay(onTime * 1000);
  digitalWrite(popPin, LOW);
}

void loop() {
}
```

❶ The constants `minDelay` and `maxDelay` specify the range of possible times after which the resistor will be powered up and the balloon might pop.

❷ The constant `onTime` specifies how long the resistor needs to be on. You may need a bit of trial and error to get this right. You don't want the resistor to be hot for too long, or it will quickly burn out.

❸ Random numbers in Arduino C are not truly random, but actually part of a long sequence. To ensure that you get a different delay each time your Arduino restarts, this line sets the position of that sequence based on the reading on analog pin A0. Since this pin is not connected to anything, the readings from it are more or less random.

❹ The pause before the resistor gets switched on is decided by the `random` function that returns a random number between the two values supplied to it as parameters. 1 is added to the second parameter because the range is exclusive. The variable used is of type `long` because an `int` can only hold numbers up to 32,767, which would stop you specifying a range of possible delays beyond 32 seconds.

❺ Delay for `pause` seconds.

❻ Turn on the transistor and power up the resistor. Delay for onTime() then turn the power off again.

Using the Balloon Popper

While you are testing the project, you might want to add a bit of certainty as to when the balloon is going to pop, so change minDelay and maxDelay to perhaps 2 and 3, respectively.

Tape the resistor to the balloon and then plug the resistor leads onto the breadboard. Click the Reset button and wait for the bang.

Heating Elements

Resistors are not really very practical as heaters. To heat anything of any size, you would use a heating element. Heating elements are really just big resistors made from materials that are designed to get hot and also to transfer that heat into whatever you are trying to make hot.

Heating things up generally requires a lot of energy in comparison with generating light, sound, or even movement. So, heating elements for things like electric kettles, domestic washing machines, and other appliances that heat water will use high-voltage AC direct from the AC line to simply get enough power to do the job. As such, when switching substantial heating elements, you will probably need to look ahead at Chapter 13.

The power of a heating element is measured in watts (W) or kW (thousands of watts) just like the power ratings of a resistor. It's really the same thing. The power rating of the resistor is just the amount of heat energy it can produce per second before it gets too hot and breaks.

In "Experiment: Resistor Heating" on page 203, we used Ohm's law and the power law (P = I V) to calculate the power. These can be combined into a single useful formula that will tell you the power if you know the resistance of the heating element and the voltage across its terminals:

$$P = \frac{V^2}{R}$$

In other words, the heat power produced by a heating element is the voltage squared in volts divided by the resistance in ohms. This formula is true for both DC and AC voltages.

As an example, a 10Ω resistor (or heating element) with 12V across it will generate (12 x 12) / 10 = 14.4 W of heat. If the voltage was 120V AC, then the power would be (120 x 120) / 10 = 1440W or 1.44 kW.

So, any heating element should have a couple things defined for it:

- Its working voltage (12V, 120V, 220V)

- Its power (50W, 1kW, 5kW)

If it's a heating element designed to heat water, it probably needs to be in water, so that the water can carry away the heat. If not, it may quickly become too hot and burn out.

Power and Energy

The words *power* and *energy* are often used interchangeably in general language. Scientifically speaking, the relationship between power and energy is as different as the relationship between speed and distance.

Power is actually the rate of conversion of energy. The power of a heating element is the amount of energy that is released as heat per second.

The unit of energy is the joule, and a watt of power is actually one joule of energy per second.

From Power to Temperature Increase

If you know how many watts of power your heating element can produce and the material it is heating, you can work out how long it will take to increase the temperature by a certain amount.

As with many things of a scientific nature, the international units of watts, degrees Celsius, grams, and seconds tend to be used for these calculations. You can always convert your answer into a unit you are more familiar with later.

Materials have a property called specific heat capacity (SHC). This is the amount of energy it takes to raise one gram of the material by 1°C. For example, water has an SHC of 4.2 J/g/degree C. In other words, it takes 4.2J of energy to raise 1 gram of water by 1°C in temperature. If you wanted to raise 100 grams of water by 1°C, it would require 420J of energy, and if you wanted to raise that same 100 grams by 10°C, the energy requirement would go up to 4200 joules.

Air has an SHC of around 1 J/g/degree C, and glass has an SCH of around 0.8 J/g/degree C. All materials are different.

Boiling Water

As an example, let's work out how well our little resistor would do at boiling some water.

Let's assume that we want to boil a cup of water (about 250g of water). Let's also assume that the water starts at room temperature (20°C) and will boil at 100°C; this means there is a temperature increase of 80°C.

The total energy needed to raise this 250g of water by 80°C is:

$$4.2 \times 250 \times 80 = 84{,}000 \text{ joules}$$

In "Experiment: Resistor Heating" on page 203, your resistor was producing 0.36W or 0.36 joules per second. So, at that rate, it would take 84,000 / 0.36 = 233,333 seconds, which is roughly 64 hours!

For practical reasons described in the next section, it's never actually going to get there.

Heat Loss and Equilibrium

It's important to make sure that your heating (or cooling) element has enough power to do the job.

The reason that you are never going to boil a cup of water with a 1/4W resistor is heat loss. If the water were in a perfectly thermally insulated container from which no heat could escape, then yes, the water would continue to get hotter and hotter until it boiled.

The loss of heat from something is proportional to how much hotter than its environment that thing is. So, as the water gets hotter, the faster it loses heat. So, eventually a balance is reached where the water container is losing heat at the same rate (watts) as the heating element is adding energy (watts) and the temperature stabilizes. This is why the resistor in "Experiment: Resistor Heating" on page 203 gets up to about 75°C and then stays there.

Peltier Elements

Peltier elements (Figure 11-5) have the very useful property that when you pass a current through them, one side of the element gets hotter and the other cooler.

You have to use a fair amount of current for this to happen (typically 2 to 6A at 12V), so to use a Peltier element you will need a fairly beefy power supply. These elements are often found in camping fridges and beverage coolers, and have the advantage over conventional refrigerators of having no moving parts to go wrong.

How Peltier Elements Work

When a current passes through a junction between two different conductive materials, one side of the junction will get slightly hotter and the other slightly cooler. This is called the *Peltier effect* after the French physicist Jean Peltier who discovered it in 1834. It is also known as the *thermoelectric effect*.

The effect is relatively small, so to make it useful enough to do something like cool down a beverage for us, the effect needs to be multiplied. This is accomplished by placing a series of alternating junctions next to one another so that current can pass through each one in series, so that each contributes to the overall effect. Common, low-cost elements typically have around 12 junctions (Figure 11-6). On either side of the elements there is a base material that forms the bread of a junction sandwich.

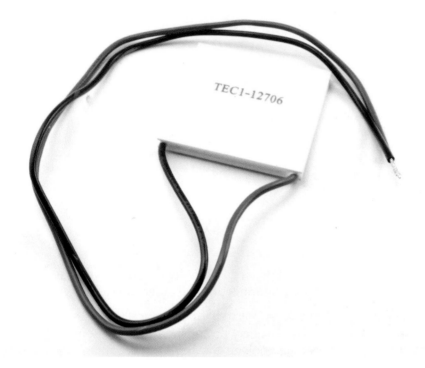

Figure 11-5 *A Peltier element*

The different types of material used in the junctions are two types of semiconductor, like those used to make transistors and chips, but optimized for their thermoelectric effect. They are known as N or P type (negative and positive).

One very interesting feature of Peltier elements is that as well as being used to cool things, if you arrange for one side to be hotter than the other a small amount of electricity is generated.

Practical Considerations

The main problem with Peltier elements is that the hot and cold sides are very close together and so the hot side will soon warm the cold side unless something is done to take the heat away as quickly as possible.

The starting point for this is to use a heat sink, as shown in Figure 11-7.

Top view

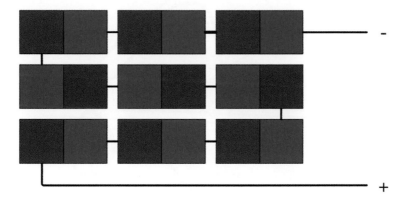

Side view   Hot Side

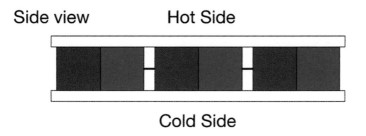

Cold Side

Figure 11-6 *Inside a Peltier element*

In the cooling unit shown in Figure 11-7, the heat sink is a lump of aluminum with blades sticking out to increase the surface area and carry away the heat. The cold side of the unit is the block arrangement that is designed to protrude into the thermally insulated refrigerated compartment.

A heat sink on its own is not nearly as effective as a heat sink with a fan attached to it to take away the air that has been warmed by convection and replace it with new cooler air. In fact, some units have a fan on both sides of the Peltier element to help it work more efficiently (Figure 11-8).

Figure 11-7 *Using a heat sink with a Peltier element*

Figure 11-8 *A Peltier refrigeration unit with dual fans*

Project: Beverage Cooler

This project does not use either a Raspberry Pi or an Arduino, but really just serves to show you how to wire up a Peltier cooling unit and how you can make yourself a simple beverage cooler (Figure 11-9). In Chapter 12, this basic project will be extended to add thermostatic control of the temperature, and then in Chapter 14, it will get an OLED display to indicate the temperature.

Figure 11-9 *The drinks cooler project*

Parts List

You are going to need the following parts to build this project:

Part	Sources
Dual-fan Peltier refrigeration unit 4A or less	eBay
Female barrel jack to screw terminal adapter	Adafruit: 368
Power supply (12V at 5A)	Adafruit: 352
Large milk or juice container	Recycling
A refreshing drink	

If you want to use a more powerful Peltier element than 4A, make sure that you also upsize your power supply to have a higher maximum current rating than the cooling unit. Allow at least an extra half amp for the fans and half an amp for luck.

Construction

If you unravel the wires of your cooling unit, you will find three pairs of wires: one pair for the Peltier element itself and one pair for each of the two fans. All of these require 12V from the power supply and the easiest way to accomplish this is to use a handy screw terminal to DC barrel socket adapter. Simply put all three red wires from the cooling unit into the screw terminal marked + and all three black wires into the screw terminal marked—as shown in Figure 11-10.

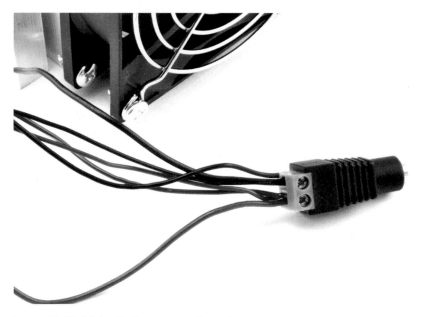

Figure 11-10 *Wiring the beverage cooler project*

You'll need to cut off the top of the plastic container and cut a square aperture into the side so that the cooling part of the Peltier cooling unit sticks into the bottle (Figure 11-11). Make sure to choose a plastic container that is large enough to contain your favorite glass or bottle for cooling.

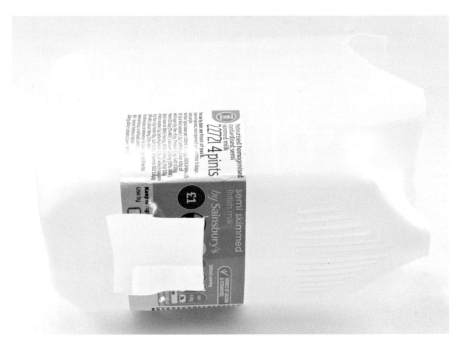

Figure 11-11 *Adapting the plastic container*

If the cool-side fan has screws that protrude, make a couple of holes in the plastic container for them too, so that the cooling unit is firmly attached to the plastic container. If you make sure that the hole for the cool-side of the Peltier unit is in the right place for the bottom of the Peltier cooler and plastic container to be level, then it won't topple over.

The great thing about using a plastic container that would otherwise be recycled is that if you make a cut in the wrong place, you can always just start over with a new container.

Using the Project

With everything assembled, plug in the power supply and both fans should start to whir. If you put your hand into the container, you should immediately feel the cold air coming from the small fan.

As we discovered with our calculations on trying to boil water, changing the temperature of even a relatively small amount of water takes a long time, so although this project will eventually cool down a warm drink, it is far better at keeping a drink cold that was cold to start with.

This project is also somewhat wasteful, as it uses up to 50W of electricity just to keep one drink cool. In Chapter 12, you can make this project a bit more efficient and add thermostatic control to it.

Summary

Heating and cooling require quite high power to work quickly, but the switching techniques using transistors and relays can all be used to switch heating and Peltier elements.

In the next chapter, you will learn how to control temperature precisely and improve the beverage cooler project to use thermostatic control.

Control Loops | 12

Although this is a book primarily about actuators (motors, heaters, lights, and other "outputs"), many systems that use actuators will also use sensors to monitor the actuator. A typical example of this is a thermostatically controlled heater, where the power to the heating element would be controlled in response to readings from a temperature sensor. This can be as simple as on/off control or a more advanced technique such as proportional-integral-derivative (PID) control.

This chapter looks at how sensors and actuators are combined to make a control system; we'll apply this technique to expand the beverage cooler project we completed in Chapter 11.

The Simple Thermostat

Perhaps the best starting point for understanding how to use a sensor and actuator in a control system is to use a temperature sensor with a heating element and attempt to keep the temperature constant. The principles learned from this can be applied equally well to controlling the position of a motor or the level of water in a tank, or any other property that can be both measured and controlled electronically.

Figure 12-1 shows why such systems are called *control loops*.

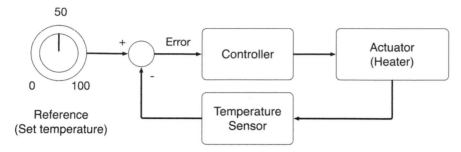

Figure 12-1 *A control loop (thermostat)*

This kind of arrangement is called a control loop. In the case of a thermostat, you set the temperature that you want your house to be at. This is the *reference* or *set temperature*. A temperature sensor measures the *actual temperature*. By subtracting the *actual temperature* from the *set temperature*, a value of *error* is found. The controller then uses this error to calculate how much power to supply to the heater.

At its simplest, the heater is just turned on or off. If this error is positive (that is, the set temperature is greater than actual temperature), it's too cold so turn the heater on. On the other hand, if actual is greater than set, it's too hot so turn it off again.

In "Experiment: How Good Is On/Off Thermostatic Control?" on page 220 you will use an Arduino with a resistor as a heater and a digital temperature sensor to make a simple thermostat. You will also use this same hardware as you learn about more accurate ways of controlling an actuator.

Experiment: How Good Is On/Off Thermostatic Control?

Our first experiment in this chapter will use an Arduino and a DS18B20 digital temperature sensor IC to implement a simple on/off temperature control system. A 100Ω resistor is used as a heating element, and this is physically attached to the DS18B20, which measures the temperature of the resistor.

Serial Monitor commands will be used to adjust the set temperature, and the temperature readings are also displayed in the Serial Monitor. You can copy and paste the temperature readings into a spreadsheet and in turn generate charts to show how well the temperature is being regulated.

By keeping things small and close together, we can experiment with temperature control without having to do things on a big (and therefore slow) scale.

Figure 12-2 shows the experiment. You can see the peg type paper clip holding the resistor firmly against the temperature sensor chip.

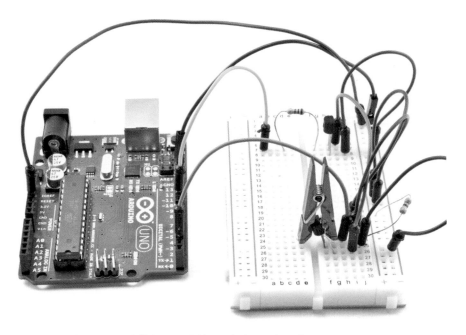

Figure 12-2 *Arduino on/off thermostatic control experiment*

Parts List

Whether you are using a Raspberry Pi or Arduino (or both), you are going to need the following parts to carry out this experiment:

Name	Part	Sources
IC1	DS18B20 digital thermometer IC	Adafruit: 374 (includes 4.7k resistor)
R1	4.7kΩ resistor	Mouser: 291-4.7k-RC
R2	1kΩ resistor	Mouser: 291-1k-RC
R3	100Ω resistor	Mouser: 291-100-RC
Q1	MPSA14 Darlington transistor	Mouser: 833-MPSA14-AP
	400-point solderless breadboard	Adafruit: 64
	Male-to-male jumper wires	Adafruit: 758
	Tiny paper peg	

If you don't have a tiny paper peg, you can use a regular paper clip, but make sure it's the plastic-coated type so that it doesn't accidentally short out the resistor or IC connections.

The DS18B20 is sometimes sold with a 4.7kΩ resistor. It is also often sold as a waterproof encapsulated package. For this experiment, the bare chip (looks like a transistor) is best, but if you want to go on and make a real thermostat, then the waterproof probe version used in "Project: A Thermostatic Beverage Cooler" on page 246 is ideal.

Design

Figure 12-3 shows the schematic diagram for the experiment.

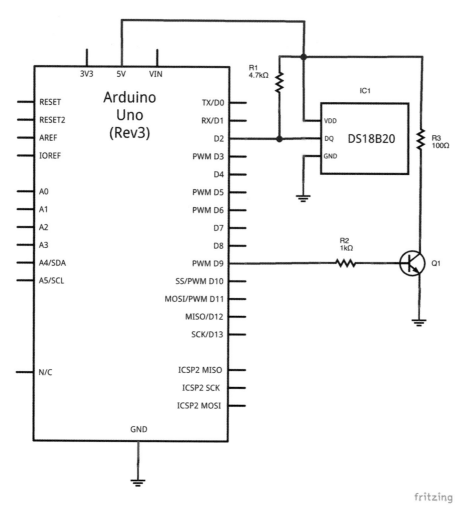

Figure 12-3 *Schematic diagram for the on/off thermostat experiment*

The schematic is really in two parts: the DS18B20 temperature sensor that needs the 4.7k pullup resistor R1 and the heating part of the circuit centered around the transistor Q1. R2 limits the current flowing into the base of the transistor and R3 is the resistor being used as a heating element.

DS18B20

The DS18B20 is a great little device. Not only is it pretty accurate (+− 0.5 degrees C), but you can also chain together a series of devices using just the one pin of your Arduino or Raspberry Pi.

You can get instructions on wiring up multiple DS18B20 sensors and addressing them individually in software with an Arduino from Miles Burton (*https://milesburton.com/Dallas_Tempera ture_Control_Library*).

When extracting the data from multiple devices on a Raspberry Pi, you have to use a different device ID when opening the device file as described in "Raspberry Pi Software" on page 241.

Breadboard Layout

Figure 12-4 shows the breadboard layout for the experiment.

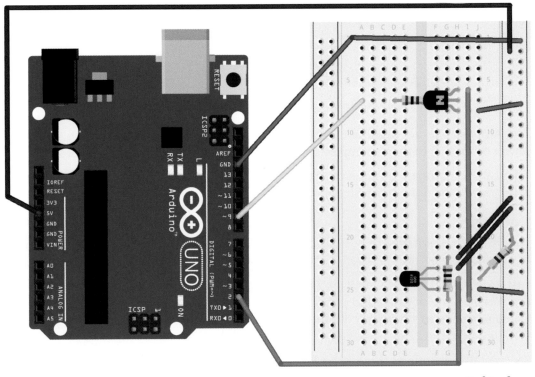

Figure 12-4 *Breadboard layout for the on/off thermostat*

The heating resistor R3 is placed right next to the flat side of the DS18B20 so that the two can be held together so that they touch using a small peg (Figure 12-5).

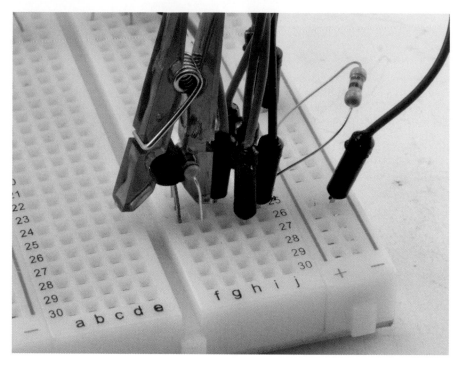

Figure 12-5 *Using a small peg to hold the resistor and DS18B20 together*

Two Arduino pins are used: pin D9 is used to control the transistor and hence switch the power on and off to the heating resistor, and pin D2 is the input from the digital sensor.

Take care to correctly position the transistor and the DS18B20 chip. Both have their flat sides toward the right of the breadboard.

Software

The DB18B20 uses a type of bus called 1-wire. To use this sensor with an Arduino, you need to download two Arduino libraries and install them into your Arduino environment.

First, download the OneWire library (*http://www.pjrc.com/teensy/arduino_libraries/ OneWire.zip*). The downloaded archive needs to be extracted and then the directory called OneWire needs to be installed into your Arduino environment (see "Installing Arduino Libraries" on page 225).

You can also download the second library for the DS18B20 itself (*https://github.com/mile sburton/Arduino-Temperature-Control-Library*). Click the Download ZIP button on the GitHub page. Once extracted, you should have a folder called *dallas-temperature-control/*, which you'll need to rename *DallasTemperature/* before moving it into your *libraries/* directory.

Installing Arduino Libraries

The Arduino IDE includes many preinstalled libraries, but from time to time you will need to download a library from the Internet that isn't included in the IDE. The procedure is pretty easy: the library is usually downloaded as a ZIP archive that extracts into a single directory. This directory name has to match the name of the library, so occasionally the directory you download will not be quite the same, especially when downloading from GitHub.

After renaming the directory (if necessary), it needs to be moved into your Arduino *libraries/* directory within your *arduino/* directory. The

Arduino IDE will automatically have created the *arduino/* directory inside *My Documents/* (on Windows) or *Documents/* on a Mac or Linux. This is where Arduino keeps your sketches.

If this is the first library that you have created, you will have to create a directory in this *arduino/* directory called *libraries/* before moving the downloaded directory into the *libraries/* directory.

You will need to exit and restart the Arduino IDE after installing a new library in order for it to be recognized.

The Arduino sketch for this project is in the *arduino/experiments/simple_thermostat/* directory (which you'll find in the place where you downloaded the book's code—see "The Book Code" on page 14 in Chapter 2):

```
#include <OneWire.h>
#include <DallasTemperature.h>

const int tempPin = 2;          ❶
const int heatPin = 9;
const long period = 1000;       ❷

OneWire oneWire(tempPin);       ❸
DallasTemperature sensors(&oneWire);

float setTemp = 0.0;            ❹
long lastSampleTime = 0;

void setup() {
  pinMode(heatPin, OUTPUT);
  Serial.begin(9600);
  Serial.println("t30 - sets the temperature to 30");
  sensors.begin();              ❺
}

void loop() {
  if (Serial.available()) {     ❻
    char command = Serial.read();
    if (command == 't') {
      setTemp = Serial.parseInt();
      Serial.print("Set Temp=");
      Serial.println(setTemp);
    }
```

```
    }
    long now = millis();              ❼
    if (now > lastSampleTime + period) {
      lastSampleTime = now;
      float measuredTemp = readTemp();     ❽
      float error = setTemp - measuredTemp;
      Serial.print(measuredTemp);
      Serial.print(", ");
      Serial.print(setTemp);
      if (error > 0) {                ❾
          digitalWrite(heatPin, HIGH);
          Serial.println(", 1");
      }
      else {
        digitalWrite(heatPin, LOW);
        Serial.println(", 0");
      }
    }
  }
}

float readTemp() {      ❿
  sensors.requestTemperatures();
  return sensors.getTempCByIndex(0);
}
```

❶ The code starts by defining constants for the temperature reading pin (tempPin) and the heater control pin (heatPin).

❷ The period constant is used to set the period between successive temperature readings. The DS18B20 is not terribly fast and can take up to 750 milliseconds, so keep the value of period above 750.

❸ Both the DallasTemperature and OneWire libraries need variables to be defined to allow access them.

❹ The variable setTemp contains the current desired temperature and the variable last SampleTime is used to keep track of when the last sample was taken.

❺ This initializes the DallasTemperature library.

❻ The first part of the loop() function checks for serial commands coming from the Serial Monitor. In fact, there is only one possible command ('t') followed by the desired set temperature in degrees Celsius. After setting the new value of setTemp, the new value is echoed back to the Serial Monitor for confirmation.

❼ The second half of loop() first checks to see if it's time to take another reading—that is, if period milliseconds have elapsed.

❽   The temperature is measured and the error calculated and then the `measuredTemp` and `setTemp` are written to the Serial Monitor.

❾   If the error is greater than 0, then turn the heater on, and end the serial monitor line with '1' to indicate power on; otherwise, turn the heater off and send a '0'.

❿   You can actually chain a whole load of DS18B20s to a single input pin, so the `reques tTemperatures` command asks all the DS18B20s connected (in this case, just one) to measure and report their temperatures. The `getTempCByIndex` function is then used to return the reading from the first and only attached DS18B20.

Experimenting

Upload the program and open the Serial Monitor. A steady stream of temperature readings should start appearing (note that the DS18B20 is designed to work in degrees Celsius and because this chapter has something of a scientific bent to it, I have stuck with temperature units of Celsius throughout):

```
t30 - sets the temperature to 30
21.75, 0.00, 0
21.69, 0.00, 0
21.75, 0.00, 0
21.69, 0.00, 0
21.75, 0.00, 0
```

The three columns are the measured temperature, the set temperature (0 at present), and whether the power is on to the heater (0 or 1).

Enter the command t30 to set the desired temperature to 30ºC. The temperature should immediately start to rise as the heater is turned on, indicated by a 1 in the third column:

```
21.75, 0.00, 0
21.75, 0.00, 0
Set Temp=30.00
21.75, 30.00, 1
21.75, 30.00, 1
21.81, 30.00, 1
21.94, 30.00, 1
22.06, 30.00, 1
```

When the temperature reaches 30º, the heat should turn off again and you should see a cycle of the heater turning on and off as the temperature oscillates close to 30º:

```
29.87, 30.00, 1
29.94, 30.00, 1
29.94, 30.00, 1
30.00, 30.00, 0
30.06, 30.00, 0
30.12, 30.00, 0
30.06, 30.00, 0
29.94, 30.00, 1
```

```
29.81, 30.00, 1
29.75, 30.00, 1
```

That's it! You have made a thermostat, and for many purposes, it will be just fine. However, you can see that the temperature "hunts" around the set temperature of 30. We can see just how much by copying a chunk of the data, pasting it into a file using a text editor, and giving it the extension *.csv*. You can then import this into your favorite spreadsheet software and create a chart from the data. If you do this, you will see something like Figure 12-6.

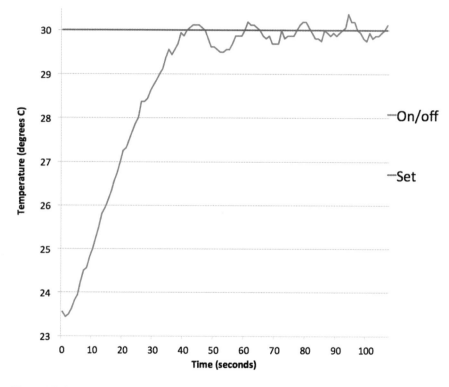

Figure 12-6 *Charting the temperature from a simple on/off thermostat*

As you can see, the temperature fluctuates by almost half a degree above and below the set temperature. We can greatly improve our results by using a control technique called proportional-integral-derivative (PID). Although PID holds the promise of being full of terrifying math, it can actually be simplified by rules of thumb.

Hysteresis

Using a transistor to switch power allows very rapid switching and the use of PWM. However, for some systems, the last thing that you need is for the system power to be switched on and off at high speed. This would be true of a domestic furnace that takes a while to start producing heat after it fires up. Also, being controlled using electromechanical valves and other parts would shorten the furnace's life by continuously turning it on and off.

To prevent too rapid switching, you could set a minimum on time and not allow the device to be turned off until that period had expired, or you could introduce hysteresis (Figure 12-7).

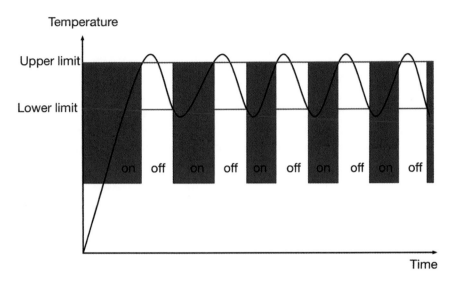

Figure 12-7 *Hysteresis in a thermostat*

In the case of a thermostat, hysteresis amounts to having two temperature thresholds rather than one, with one threshold being a fixed amount above the other. If the temperature falls below the lower threshold, then the heater is turned on, but it is only turned off again when it exceeds the higher threshold.

In this way, the natural inertia of the system is used to introduce delays into the switching.

PID Control

Turning the heater off when the temperature sensor indicates that it's hotter than the set temperature and on when it's below the set temperature results in a system where the temperature "hunts" around the set temperature, as shown in Figure 12-6.

If you need the temperature of your system to be maintained more accurately, then you need to use proportional-integral-derivative (PID) control.

Rather than simply switch a heater on and off, a PID controller varies the output power to the heater (or other actuator) taking three factors into consideration. These factors are as follows: proportional, integral, and derivative.

Warning: This section gets a bit deep. PID control is something that I get asked about a lot and one of the main goals of this chapter is to explain how PID control works and how to use it.

Proportional (P)

You can get pretty good results in many systems using just the P part of PID controller software and ignoring the I and D parts. In a new system, you should start with just the P part anyway and see how it goes.

Proportional control just means that the output power to the heater is proportional to the error. So, the bigger the error—that is, the further the actual temperature is from the set temperature—the more power goes to the heater. As the actual temperature gets close to the set temperature, the power eases off so that the temperature does not overshoot as much as it did in Figure 12-6. Depending on the system, it will probably still overshoot, but not nearly as much as it does with a simple on/off control. It's a bit like driving up to a stop sign: you should anticipate the stop sign and brake before you reach it rather than slam on the brakes once you're there.

If the actual temperature is higher than the set temperature, then the error will be negative, resulting in the need for a negative value of power to the heater. If the heater was a Peltier element (Chapter 11), you could at this point reverse the current through it (using an H-bridge) and start cooling rather than heating. In practice, you don't normally need to do this, unless the ambient temperature is very close to the set temperature. Instead, you can just let the system cool (or heat) naturally.

The error on which we base the calculation for the output power is, in the case of a thermostat, a temperature difference. This might be in degrees Celsius, so if the set temperature is 30°C and the measured temperature is 25°C, then the error is 5°C (30°C – 25C°). If you are using PWM of an Arduino to set the output power, then the output needs a value between 0° and 255°C. So, directly setting the output to the error (5) would deliver very little power to the heater; in fact, probably not enough to ever get the temperature up to 30°C. For this reason, the error is multiplied by a number called kp (also some times called *gain*) to give the output power. Changing kp will determine how fast the temperature will get up to the set temperature. With a low value of kp the temperature may never get up to the set temperature, but if the kp value is too high, the system will behave exactly like the on/off temperature controller and oscillate back and forth about the set temperature. Figure 12-8 shows how different values of kp alter how an idealized system behaves.

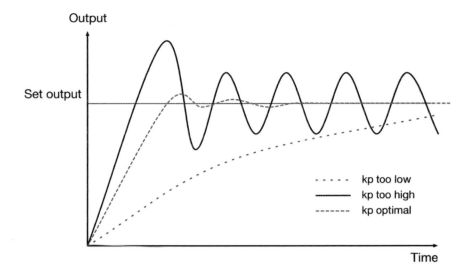

Figure 12-8 *The effect of kp (gain) on output in a proportional controller*

If kp is not large enough, then the set output may never be reached or may take a very long time to get there. On the other hand, if kp is too high the output will oscillate and the magnitude of those oscillations will not decrease. A desirable value of kp will quickly raise the output to the set level, maybe overshoot a little, and then the oscillations will quickly die down (settling) to an acceptably low level.

Finding the right value of kp is called tuning the system. In "Tuning a PID Controller" on page 232 you will learn about tuning a PID control system.

Let's return to the example of a set temperature of 30ºC but an actual temperature of 25ºC. At this point, we still want maximum power to go to the heater—that is, a PWM duty of 255 (100%). If we picked a value of kp of 50, then the error of 5 would result in an output power of:

$$\text{Output} = \text{error} * kp = 5 * 50 = 250$$

Once the system was only 1º away from the set temperature, the output power would fall to:

$$\text{Output} = \text{error} * kp = 1 * 50 = 50$$

That may or may not be enough power to allow the system to reach the set temperature. Every system is different, hence the need to tune the controller.

Integral (I)

Assuming that proportional power control of the output is not accurate enough, you may need to add some I to the calculation, so that the output power is calculated as:

$$\text{Output} = \text{error} * kp + I * ki$$

There is a new constant ki that will scale this mysterious new property called I (integral) and the output power will be calculated by adding this integral term to the proportional term. This type of controller is called a PI (or sometime P + I) controller (not to be confused with Pi of Raspberry Pi). Just like proportional-only controllers, sometimes PI will produce good enough results, without having to add D.

The integral term is calculated by keeping a running total of the errors, every time the temperature is measured. While the error is positive (warming up), the I term will get bigger and bigger, boosting the initial response. It will only actually start to decrease once the error becomes negative, when the set temperature has been exceeded.

Once the actual temperature gets to the set temperature, the I term has a calming effect, smoothing the temperature changes so that the temperature settles better to the desired value.

The only problem with the I term is that because it gives a big boost as the temperature is rising, this leads to what can be a big overshoot of the temperature that then takes a while to settle back down to the set temperature and stabilize.

Derivative (D)

In practice, the D term is often not used in real control systems, because its advantages in reducing overshoot are outweighed by making tuning more difficult.

To counteract the overshoot effect, a D term can be added to the control software, so that now, the output power is calculated as follows:

$$Output = error * kp + I * ki + D * kd$$

This D term is a measure of how fast the error is changing between each measurement of the temperature, and so in a sense predicts where the temperature is going.

Tuning a PID Controller

Tuning a PID controller means finding the values of kp, ki, and kd that make your system behave the way that you want it to behave. I would recommend that you make life easy for yourself and just use PI control, setting kd to 0. This just leaves you two parameters to tune. In fact, this simplification makes most systems easy but often time consuming to tune.

In "Experiment: PID Thermostatic Control" on page 233, I will take you through a trial-and-error approach that works fine for the kind of temperature controller used. The remainder of this section is devoted to one of the most popular techniques for PID tuning called the *Ziegler-Nichols method*. This essentially reduces the process to a series of experiments and then a few simple calculations to give the values of kp, ki, and kd.

The Ziegler-Nichols tuning starts by setting ki and kd to 0 so that the controller is operating in proportional mode. The value of kp is then steadily increased until the output starts to oscillate. The value of kp at which this happens is called ku.

Having found ku, you then need to measure the period of oscillation in seconds (Figure 12-9); this is called pu.

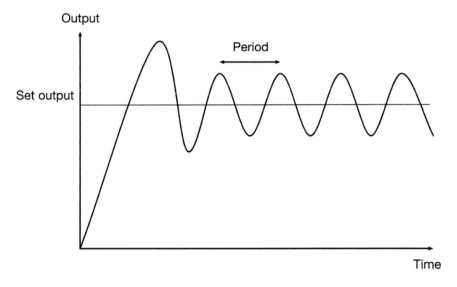

Figure 12-9 *Finding the period of oscillation*

To do this, you will need to plot a load of readings like you did in "Experiment: How Good Is On/Off Thermostatic Control?" on page 220.

You can then calculate the values for kp, ki, and kd as follows:

$$kp = 0.6 * ku$$
$$ki = (2 * kp) / pu$$
$$kd = (kp * pu) / 8$$

If your controller is just PI, then the following calculations are recommended:

$$kp = 0.45 * ku$$
$$ki = (1.2 * kp) / pu$$
$$kd = 0$$

You can find chapter and verse on using the Ziegler-Nichols method on Wikipedia (*https://en.wikipedia.org/wiki/PID_controller*).

Experiment: PID Thermostatic Control

In this experiment, you can try out PID control using both an Arduino and a Raspberry Pi. The Arduino PID controller software is from a library, but the PID control software for the Raspberry Pi version is written by hand.

Hardware

This experiment uses the exact same hardware as "Experiment: How Good Is On/Off Thermostatic Control?" on page 220. However, care must be taken when connecting it to the Raspberry Pi, as the DS18B20 and the heating resistor need to be powered separately on the Pi; the DS18B20 from 3.3V and the heating resistor from 5V. This is because the digital output from the DS18B20 needs to be pulled up to 3.3V rather than 5V so that it does not damage the Raspberry Pi's input pin.

Arduino Software

This experiement uses a ready-made PID library (*https://github.com/br3ttb/Arduino-PID-Library/*) that you can download and that needs and instal as described in "Installing Arduino Libraries" on page 225. You can find full documentation for this library at the Arduino website (*http://playground.arduino.cc/Code/PIDLibrary*).

The Arduino sketch for the experiment can be found in */arduino/experiments/pid_thermostat*:

```
#include <OneWire.h>
#include <DallasTemperature.h>
#include <PID_v1.h>

const int tempPin = 2;
const int heatPin = 9;
const long period = 1000; // >750

double kp = 0.0;          ❶
double kd = 0.0;
double ki = 0.0;

OneWire oneWire(tempPin);
DallasTemperature sensors(&oneWire);

double setTemp = 0.0;
double measuredTemp = 0.0;
double outputPower = 0.0;      ❷
long lastSampleTime = 0;

PID myPID(&measuredTemp, &outputPower, &setTemp, kp, ki, kd, DIRECT); // ❸

void setup() {
  pinMode(heatPin, OUTPUT);
  Serial.begin(9600);
  Serial.println("t30 - sets the temperature to 30");
  Serial.println("k50 20 10 - sets kp, ki and kd respectively");
  sensors.begin();
  myPID.SetSampleTime(1000);   ❹
  myPID.SetMode(AUTOMATIC);
}

void loop() {
  checkForSerialCommands();     ❺
```

```
  long now = millis();
  if (now > lastSampleTime + period) {    ❻
    lastSampleTime = now;
    measuredTemp = readTemp();
    myPID.Compute();
    analogWrite(heatPin, outputPower);

    Serial.print(measuredTemp);    ❼
    Serial.print(", ");
    Serial.print(setTemp);
    Serial.print(", ");
    Serial.println(outputPower);
  }
}

void checkForSerialCommands() {    ❽
    if (Serial.available()) {
    char command = Serial.read();
    if (command == 't') {
      setTemp = Serial.parseFloat();
      Serial.print("Set Temp=");
      Serial.println(setTemp);
    }
    if (command == 'k') {
      kp = Serial.parseFloat();
      ki = Serial.parseFloat();
      kd = Serial.parseFloat();
      myPID.SetTunings(kp, ki, kd);
      Serial.print("Set Constants kp=");
      Serial.print(kp);
      Serial.print(" ki=");
      Serial.print(ki);
      Serial.print(" kd=");
      Serial.println(kd);
    }
  }
}

double readTemp() {
  sensors.requestTemperatures();
  return sensors.getTempCByIndex(0);
}
```

The code concerned with getting a temperature reading out of the DS18B20 is the same as for "Experiment: How Good Is On/Off Thermostatic Control?" on page 220 so I will concentrate on the code that relates to PID control.

❶   Three variables (kp, ki, and kd) are defined as the scaling factors for the P, I, and D components of the output. Variables are used because these can be changed using a Serial Monitor command while the program is running. The type double is used rather than `float`, in part because they are more precise than `float`s, but also because that's what the library expects.

❷ The variable `outputPower` will contain the PWM duty cycle for the heater of between 0 and 255.

❸ To access the PID library code, a variable is defined called `myPID`. You will notice that the first three parameters when creating a PID variable are the names of the variables `measuredTemp`, `outputPower`, and `setTemp` prefixed by an `&`. This is a C trick that allows the PID library to modify the values in those variables even though the variables are not part of the library. If you want to learn more about this technique (C pointers), take a look at Tutorials Point (*http://www.tutorialspoint.com/cprogramming/c_point ers.htm*). The final parameter (`DIRECT`) sets the mode of PID operation to be direct, which in the case of this library means that the output will be proportional to the error, rather than inverted. Usefully, by default, this library scales the output to 0 to 255 for PWM.

❹ The sample time needs to be set to 1 second (1,000 milliseconds). Setting the mode to `AUTOMATIC` starts the PID calculations.

❺ Checking for serial commands is now in its own function to stop `loop()` getting too long-winded to read easily. See also callout (8).

❻ If it's time for another sample, the temperature is read into the `measuredTemp` variable and then the PID library told to update its calculations (`myPID.Compute`). This automatically updates the value of `outputPower`, which is then used to set the PWM duty cycle of the pin used to control the heating resistor.

❼ The values are all printed back to the Serial Monitor, as we will need them to plot some charts and see how the controller is performing.

❽ The `checkForSerialCommands` function checks for a `'t'` command to set the temperature just like "Experiment: How Good Is On/Off Thermostatic Control?" on page 220 but also checks for a k command followed by three numbers (`kp`, `ki`, and `kd`) setting these tunings if the command is received.

Arduino Experimentation

The hardware and software for this experiment give us everything we need to experiment with a PID controller. We can change the set temperature and the three values of `kp`, `ki`, and `kd` and record what effect they have on the output. In this case, we will get good enough results with PI control, so `kd` can just be set to 0.

Upload the sketch to the Arduino and open the Serial Monitor (Figure 12-10).

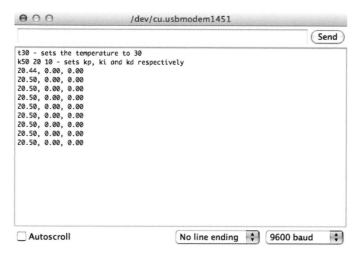

Figure 12-10 *Using the Serial Monitor to test PID control*

Tuning the controller is going to take a while. You need to record data and plot it to see how well the system behaves. The starting point is to find a good value of kp. Let's just try values of 50, 500, and 5000 to get an idea of the system.

First set the tuning parameters by entering the following into the Serial Monitor:

```
k50 0 0
```

This sets kp to 50 and ki and kd to 0. If you prefer, you can enter the values with a decimal part (for example, k30.0 0.0 0.0). In both cases, the numbers will be converted to a float.

Now let's set the desired temperature to 30°C (I chose this as a convenient temperature about 7° or 8° above the ambient temperature):

```
t30
```

The temperature should start to climb with the three columns displaying the actual temperature, the set temperature, and the PWM output value (0 to 255). Here is a sample of the data. Note that the PWM output is stored as a float and therefore has numbers after the decimal point. This will be truncated to a whole number between 0 and 255 when the analogWrite function is called:

```
25.06, 30.00, 246.88
25.19, 30.00, 240.63
25.31, 30.00, 234.38
25.44, 30.00, 228.13
```

The last column, the PWM value, is decreasing almost immediately with kp set to 50. This means 50 is much too low, but carry on gathering data for a few minutes. Then paste the data into a text file and import it into your spreadsheet software. I started copying the data when the temperature got up to 25°C. Plot the temperature and you should get a chart something like Figure 12-11.

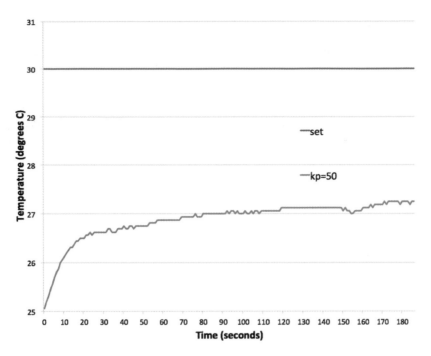

Figure 12-11 *Temperature against time for kp = 50*

Importing Data into a Spreadsheet

Getting the raw data from the Serial Monitor into a spreadsheet and then using the spreadsheet's chart drawing feature to see what is happening in the data is a very useful technique.

If you are using OpenOffice, then select the data from the Serial Monitor and type CTRL-C on Windows and Linux or Command-C on a Mac to copy the data into the clipboard. Switch over to an OpenOffice spreadsheet document, select the cell where you want the data to go, and type CTRL-V on Windows and Linux or Command-V on a Mac.

Because the data has multiple columns, OpenOffice will display the dialog shown in Figure 12-12 to try separating the columns.

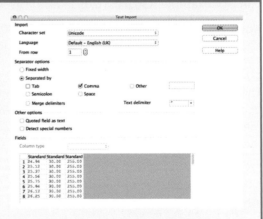

Figure 12-12 *Importing data into OpenOffice*

Select the Comma option in the "Separated by" section, hit OK, and the data will be pasted into separate columns in your spreadsheet.

If you are using Microsoft Excel, you will need to use a text editor such as Notepad++ or Textmate. First, paste the data into a new text document and then save that file with the extension *.csv* (comma separated values). You will then be able to open the file directly with Excel. You can also use the Excel Import command to import the file.

With a kp of 50, it seems unlikely that the temperature is ever going to get up to 30ºC. Set the temperature to 0 (t0 in the Serial Monitor) and wait for the system to cool back down and then repeat the procedure first for a kp of 500 and then 5000. The results of all three values of kp are shown in Figure 12-13.

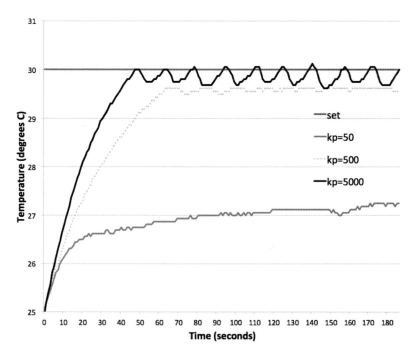

Figure 12-13 *Temperature against time for three values of kp*

As we suspected, a kp of 50 is much too low, 500 is not bad, but it does not quite reach the set temperature, and 5000 is behaving just like an on/off thermostat. Just as a guess, it looks like 700 might make a good value of kp, especially if it has a bit of a boost getting it up to the set temperature by adding a bit of I.

The Ziegler-Nichols method of tuning suggests a value of ki for a PI controller of:

$$ki = (1.2 * kp) / pu$$

We have estimated a kp of 700 and from Figure 12-13 pu is about 15 seconds, which would suggest a ki of 56.

Record another set of data with kp=700 and ki=56. The results of this are shown in Figure 12-14 along with the results for proportional-only control with kp set to 700.

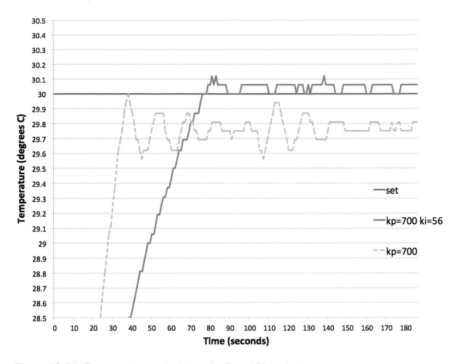

Figure 12-14 *Temperature against time for P and PI control*

Figure 12-14 is zoomed right in on the Y-axis, so that we can see just how good the results are for the PI control.

Once the PI plot reaches the set temperature, it only varies by a little over 0.1 of a degree. The temperature changes in steps rather than smoothly, because the DS18B20 is a digital device with a fixed precision.

With more effort and experimentation, it would probably be possible to improve on this result, but frankly, it's not worth the bother.

Connecting the Raspberry Pi

The breadboard layout for the Raspberry Pi (Figure 12-15) is slightly different from the Arduino. The difference is that although we still want the heating resistor to operate at 5V, the DS18B20 temperature sensor needs to operate at 3.3V to be compatible with the Raspberry Pi's 3.3V GPIO pins.

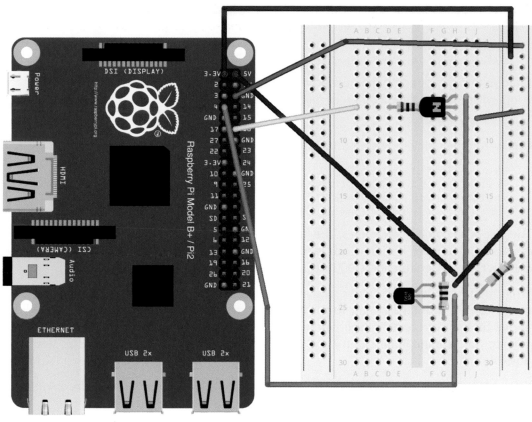

Figure 12-15 *Breadboard layout and Raspberry Pi connections*

Raspberry Pi Software

Getting the DS18B20 to work with Raspberry Pi takes a little preparation. The first step is to enable the 1-wire bus. To do this, edit the file */boot/config.txt* using the command:

```
$ sudo nano /boot/config.txt
```

Add the following line to the end of the file:

```
dtoverlay=w1-gpio
```

You now need to reboot your Raspberry Pi for the changes to take effect. The DS18B20 uses a text file–style interface, which means the Python program will need to read a file and then extract the temperature measurement from it. You can try this out before running the full program and see what the message format is by changing directory to */sys/bus/w1/devices* using this command:

```
$ cd /sys/bus/w1/devices
```

You then need to list the directories in this folder using:

```
$ ls
28-000002ecba60  w1_bus_master1
pi@raspberrypi /sys/bus/w1/devices $
```

Change into the directory name starting with 28. In this case, *28-000002ecba60* (note that yours will probably have a different name):

```
$ cd 28-000002ecba60
```

Finally, run the following command to fetch the last temperature reading:

```
$ cat w1_slave
53 01 4b 46 7f ff 0d 10 e9 : crc=e9 YES
53 01 4b 46 7f ff 0d 10 e9 t=21187
pi@raspberrypi /sys/bus/w1/devices/28-000002ecba60 $
```

The response comes as two lines. The first part of each line is the unique ID for the temperature sensor and the first line ends in YES, indicating a successful reading. The second line ends in the temperature in 1/1000 degrees C. So in this case, 21187 /(or 21.187ºC).

While there are Python libraries available for PID control, they are not as easy to use as their Arduino counterpart and so for the Raspberry Pi version, the PID algorithm will be implemented from scratch (well, not quite from scratch—the code was written with reference to the Arduino library, to try and keep the behavior of the two versions as similar as possible).

You can find the code in the file *pid_thermostat.py* (located in the *python/experiments/* directory):

```
import os
import glob
import time
import RPi.GPIO as GPIO

GPIO.setmode(GPIO.BCM)

heat_pin = 18
base_dir = '/sys/bus/w1/devices/'    ❶
device_folder = glob.glob(base_dir + '28*')[0]
device_file = device_folder + '/w1_slave'

GPIO.setup(heat_pin, GPIO.OUT)
heat_pwm = GPIO.PWM(heat_pin, 500)
heat_pwm.start(0)

old_error = 0    ❷
old_time = 0
measured_temp = 0
p_term = 0
i_term = 0
d_term = 0
```

```python
def read_temp_raw():       ❸
    f = open(device_file, 'r')
    lines = f.readlines()
    f.close()
    return lines

def read_temp():       ❹
    lines = read_temp_raw()
    while lines[0].strip()[-3:] != 'YES':
        time.sleep(0.2)
        lines = read_temp_raw()
    equals_pos = lines[1].find('t=')
    if equals_pos != -1:
        temp_string = lines[1][equals_pos+2:]
        temp_c = float(temp_string) / 1000.0
        return temp_c

def constrain(value, min, max):       ❺
    if value < min :
        return min
    if value > max :
        return max
    else:
        return value

def update_pid():       ❻
    global old_time, old_error, measured_temp, set_temp
    global p_term, i_term, d_term
    now = time.time()
    dt = now - old_time       ❼
    error = set_temp - measured_temp # ❽
    de = error - old_error       # ❾

    p_term = kp * error                       ❿
    i_term += ki * error                      ⓫
    i_term = constrain(i_term, 0, 100)        ⓬
    d_term = (de / dt) * kd                   ⓭

    old_error = error
    # print((measured_temp, p_term, i_term, d_term))
    output = p_term + i_term + d_term         ⓮
    output = constrain(output, 0, 100)
    return output

set_temp = input('Enter set temperature in C ')  # ⓯
kp = input('kp: ')
ki = input('ki: ')
kd = input('kd: ')

old_time = time.time()       ⓰
try:
    while True:
        now = time.time()
```

```
        if  now > old_time + 1 :   ⑰
            old_time = now
            measured_temp = read_temp()
            duty = update_pid()
            heat_pwm.ChangeDutyCycle(duty)

            print(str(measured_temp) + ', ' + str(set_temp) + ', ' + str(duty))
    finally:
        GPIO.cleanup()
```

❶ This code determines the directory where the file for the DS18B20 is located. It does this in much the same way as the method described earlier, using the `glob` command to find the first directory starting with 28.

❷ These global variables are used by the PID algorithm. The variable `old_error` is used to calculate the change in error for the D term.

❸ The function `read_temp_raw` reads the DS18B20 as two lines of text.

❹ The `read_temp` function is responsible for actually extracting the temperature from the end of the second line, after checking that we got the response YES on the first line.

❺ This utility function constrains the value of the first parameter so that it always lies between the range specified in the second and third parameters.

❻ The `update_pid` function is where the actual PID calculation occurs.

❼ Calculate `dt` (how much time has elapsed since the last time `update_pid` was called).

❽ Calculate the error.

❾ Calculate the change in error `de`.

❿ Calculate the proportional term.

⓫ Add the current `error * ki` to the `i_term`.

⓬ Constrain the `i_term` to lie within the same range as the output (0 to 100).

⓭ Calculate the `d_term`.

⓮ Add all the terms together and then constrain them to the output range of 0 to 100.

⓯ Unlike the Arduino version, which allows the tuning variables to be adjusted while the controller is running, the Python program just prompts you once for the temperature, kp, ki, and kd.

⑯   The old_time is initialized to the current time just before the main control loop starts.

⑰   If 1 second has elapsed since the last sample, measure the temperature and then get the new value of output (duty) and change the PWM channel's duty cycle accordingly.

Raspberry Pi Experimentation

One difference between the Python and Raspberry Pi versions is that the Raspberry Pi has an output of 0 to 100, whereas the Arduino's output is 0 to 255. So, the parameters kp and ki that we found when tuning the Arduino setup need to be adjusted for the Raspberry Pi. In fact, you can just divide kp and ki by 2.5 to put the output roughly in the range 0 to 100. This results in a value for kp of 280 and ki of 22.

Run the program, set the temperature to 30, and plug in these numbers and you should get similar results to the Arduino version:

```
$ sudo python ex_11_pid_thermostat.py
Enter set temperature in C 30
kp: 280
ki: 22
kd: 0
23.437, 30, 100
23.437, 30, 100
23.5, 30, 100
23.562, 30, 100
23.687, 30, 100
```

Charting these results using a spreadsheet, I got the results shown in Figure 12-16. Again, this is a zoomed-in view of the temperature, which is being regulated pretty accurately.

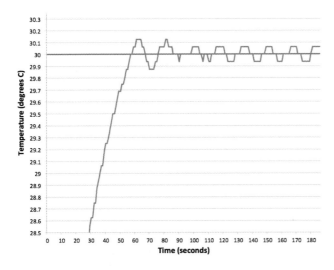

Figure 12-16 *The results of PID with Raspberry Pi*

Project: A Thermostatic Beverage Cooler

In this project, we'll add thermostatic control to the beverage cooler you made in "Project: Beverage Cooler" on page 215 so that your beverage can be cooled to just the right temperature (Figure 12-17). In Chapter 14, the project will be extended even further to add a display that shows the set and actual temperatures.

Figure 12-17 *A thermostatic drinks cooler*

This project is built with an Arduino, but given what you have learned about using Raspberry Pi with the DS18B20, you should have no problem altering the project to work with a Raspberry Pi.

Hardware

This project is built on "Project: Beverage Cooler" on page 215 but adds an Arduino and DS18B20 temperature sensor into the loop so if you have not already done so, follow the instructions for creating "Project: Beverage Cooler" on page 215.

Parts List

You are going to need the following parts to build this project:

Name	Part	Sources
R1	4.7kΩ resistor	Mouser: 291-4.7k-RC
R2	1kΩ resistor	Mouser: 291-1k-RC
R3	270Ω resistor	Mouser: 291-270-RC

R4	10kΩ Trimpot	Adafruit: 356
		Sparkfun: COM-09806
	Encapsulated DS18B20 temperature probe	eBay, Adafruit: 381
Q1	FQP30N06L MOSFET	Mouser: 512-FQP30N06L
LED1	Green LED	Adafruit: 298
		Sparkfun: COM-09650
	Dual-fan Peltier refrigeration unit 4A or less	eBay
	Female barrel jack to screw terminal adapter	Adafruit: 368
	Power supply (12V at 5A)	Adafruit: 352
	Two-way terminal block	Electrical/DIY store
	Large milk or juice container	Recycling

The encapsulated DS18B20 probe contains the same chip as you used in "Experiment: How Good Is On/Off Thermostatic Control?" on page 220 and "Experiment: PID Thermostatic Control" on page 233 except that it comes in a handy waterproof capsule with a long lead that can connect to the breadboard.

If you want to use a more powerful Peltier element than 4A, then make sure that you also upsize your power supply to have a higher maximum current rating than the cooling unit. Allow at least an extra half amp for the fans and half an amp for luck.

Design

Figure 12-18 shows the schematic diagram for this project. On the left of Figure 12-18, you can see R4, which is a variable resistor, also known as a pot (see "Potentiometers" on page 247). The slider of the pot is connected to the A0, analog input of the Arduino (see "Analog Inputs" on page 18). The position of the pot's knob sets the voltage at A0 which the Arduino measures and then uses to set the desired temperature for the cooler.

Potentiometers

A *potentiometer* or more commonly *pot* will be familiar to you as the volume control knob of a radio or amplifier. It has a knob that rotates through almost a whole revolution.

The area of Figure 12-18 around R4 shows how the pot is used as an input to an Arduino. The top of the pot is connected to 5V and the bottom of the pot is connected to ground. The middle connection of the pot will then vary between 0 and 5V depending on the position of the knob.

The right of Figure 12-18 is very similar to the schematic for "Experiment: How Good Is On/Off Thermostatic Control?" on page 220 except that instead of using a low-power MPSA14 transistor a high-power FQP30N06L MOSFET is used. This transistor can switch the 4A or more current to the cooling unit without getting warm enough to even need a heat sink.

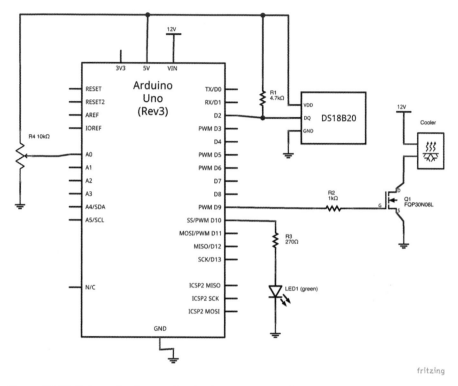

Figure 12-18 *Schematic diagram for the thermostatic cooler*

Construction

Assuming that you have built "Project: Beverage Cooler" on page 215, these are the additional steps that you need to take to build the project.

Step 1: Add the temperature probe

The physical construction of the cooling unit is just the same as "Project: Beverage Cooler" on page 215 with the addition of the temperature probe that sits at the base of the container, with the glass or bottle to be cooled sitting on top of it (Figure 12-19). In this case, I just taped the probe down, but it would be better to glue it into place.

Figure 12-19 *Adding the temperature probe*

Step 2: Construct the breadboard

Figure 12-20 shows the breadboard layout for the project and also how the various parts of the project are connected.

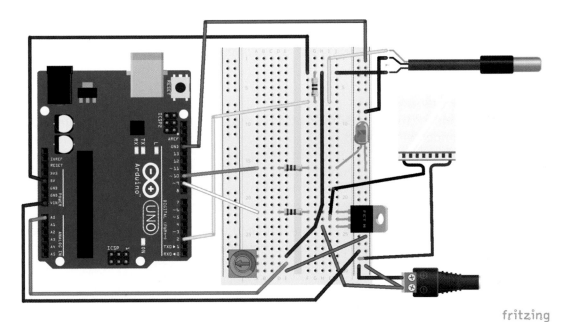

fritzing

Figure 12-20 *The breadboard layout for the project*

Add the components to the breadboard, making sure that the MOSFET and LED are positioned correctly. The temperature probe has four wires in its lead. The red and black leads are the VCC and GND connections and the yellow lead is the digital output of the probe. The fourth wire should not be connected.

I poked the trimpot leads through a small piece of paper, allowing a rudimentary scale to be devised so that you can see the temperature that the cooler is being set to.

Step 3: Attach the cooling unit

The cooling unit actually has three pairs of leads: two pairs for the fans and one for the Peltier unit itself. To make it a bit easier to connect up, a two-way terminal block is used to connect the cooling unit (Figure 12-21), allowing just two wires to connect the unit to the breadboard.

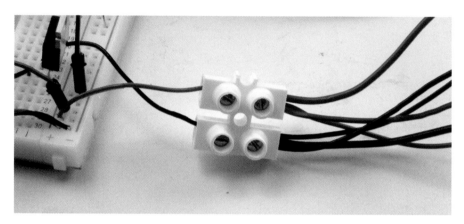

Figure 12-21 *Attaching the cooling unit*

Step 4: Attach the power socket

The female barrel jack to screw terminal adapter can be linked to the breadboard using male-to-male jumper wires. This is fine if the jumper wires are high quality and use relatively thick wire; however, many jumper wires use quite thin conductors that will get pretty warm with a few amps flowing through them. This is not in itself a problem unless the wires get hot rather than warm. It does, however, mean that not all of the 12V will be finding its way to the Peltier element and the cooler will take longer to get cool.

Alternatively, you can use some single-core insulated wire to connect the breadboard to the barrel jack adapter. The same applies to the jumper wires to the cooling unit.

Arduino Software

Using PID control for a beverage cooler could definitely be considered overkill. However, it's just a matter of software, so it won't cost any more to have a kick-ass control algorithm keeping our beverages cool.

The sketch has a lot of similarities with "Experiment: How Good Is On/Off Thermostatic Control?" on page 220 and "Experiment: PID Thermostatic Control" on page 233, including all the code for interfacing with the DS18B20 temperature sensor, so refer back to those projects for information on that part of the software:

```
#include <OneWire.h>
#include <DallasTemperature.h>
#include <PID_v1.h>

const double minTemp = 0.0;    ❶
const double maxTemp = 20.0;
const float tempOKMargin = 0.5;

double kp = 1500;    // ❷
double ki = 50.0;
double kd = 0.0;

const int tempPin = 2;
const int coolPin = 9;
const int ledPin = 10;    // ❸
const int potPin = A0;
const long period = 1000; // >750

OneWire oneWire(tempPin);
DallasTemperature sensors(&oneWire);

double setTemp = 0.0;
double measuredTemp = 0.0;
double outputPower = 0.0;
long lastSampleTime = 0;

PID myPID(&measuredTemp, &outputPower,
        &setTemp, kp, ki, kd, REVERSE); ❹

void setup() {
  pinMode(coolPin, OUTPUT);
  pinMode(ledPin, OUTPUT);
  Serial.begin(9600);
  sensors.begin();
  myPID.SetSampleTime(1000);
  myPID.SetMode(AUTOMATIC);
}

void loop() {    // ❺
  long now = millis();
  if (now > lastSampleTime + period) {
      checkTemperature();
      lastSampleTime = now;
  }
  setTemp = readSetTempFromPot();    // ❻
}

void checkTemperature() {    // ❼
```

```
    measuredTemp = readTemp();
    Serial.print(measuredTemp);
    Serial.print(", ");
    Serial.print(setTemp);
    Serial.print(", ");
    Serial.println(outputPower);
    myPID.Compute();
    analogWrite(coolPin, outputPower);
    float error = setTemp - measuredTemp; // ❽
    if (abs(error) < tempOKMargin) {
      digitalWrite(ledPin, HIGH);
    }
    else {
      digitalWrite(ledPin, LOW);
    }
  }

  double readSetTempFromPot() {    // ❾
    int raw = analogRead(potPin);
    double temp = map(raw, 0, 1023, minTemp, maxTemp);
    return temp;
  }

  double readTemp() {
    sensors.requestTemperatures();
    return sensors.getTempCByIndex(0);
  }
```

❶   The two constants `minTemp` and `maxTemp` set the range of temperatures that can be set using the pot. The `tempOKMargin` variable determines the amount above or below the set temperature the actual temperature can be before the green LED turns off.

❷   kp is set to a high value so that the heater is turned on and off fairly cleanly. This is mostly to stop the whiney noise that the fan motors make when powered under a low output. An alternative to this would be to connect the fans separately so that they are running all the time and to just control the power to the Peltier element.

❸   Extra pins are defined for the `led` and `pot`.

❹   The PID is initialized with a mode of `REVERSE` rather than `DIRECT` (as previously), because adding more output power will reduce the temperature not increase it.

❺   The main loop checks to see that a second has elapsed and then calls `checkTempera ture` to turn the cooler on or off as needed.

❻   Every time around the loop (which is going to be many times per second), the `read SetTempFromPot` function is called to set the `setTemp` variable from the position of the pot.

❼ checkTemperature measures the temperature, reads the temperature, and then updates the PID controller. This function also writes out the readings to the Serial Monitor, so that you can tune the cooler or monitor its performance. The Arduino does not need to be connected via USB, because it receives its power via its Vin pin, but if you do connect it by USB, you can see this output over the Serial Monitor.

❽ The remainder of this function turns on the LED if the measured temperature is within tempOKMargin of the set temperature. The abs (absolute) function effectively removes any minus sign from the front of a number.

❾ The code to turn the pot position into a value between minTemp and maxTemp; the raw analog reading (between 0 and 1023) is read into the variable raw. The map function (see "The Arduino map Function" on page 253) is then used to convert this to the desired temperature range.

The Arduino map Function

When using Arduino or Raspberry Pi to control something, a common problem you'll run into is converting a number that has one range of values into a number within some other range.

For example, an Arduino analog input has a range of 0 to 1023, and if we wanted to map that range onto a temperature between 0 and 20, we could just divide the number by 51.15 (1023 / 20). So that 1023 would become 1023/51.15=20.

This is not quite so easy if the ranges do not both start at 0. That is where the Arduino map function comes in. As shown here, it takes five parameters that will convert a number in the range 0 to 1023 to one in the range 20 to 40:

```
map(value, 0, 1023, 20, 40);
```

The first parameter is the value to be converted, the second and third are the range of the number you have, and the fourth and fifth are the range you would like the number to be in, in this case between 20 and 40.

Python does not have a built-in range function, but it is easy enough to write one and then use it in your programs. It goes something like this:

```
def map(value, from_low, from_high,
to_low, to_high):
        from_range = from_high -
from_low
    to_range = to_high - to_low
    scale_factor = from_range / to_range
    return to_low + (value /
scale_factor)
```

You can then call this Python function with the same parameters as you would its Arduino counterpart. For example:

```
map(510, 0, 1023, 20, 40)
```

This would return a value of 30, which is half way between 20 and 40, just as 510 is roughly half way between 0 and 1023.

Summary

Although this chapter has used temperature to illustrate how you can control something accurately as long as you can measure it and alter it, the same principles also apply for controlling other things such as position. This is, in fact, how a servomotor (Chapter 9) works.

In the next chapter we will look at how you can safely control high-voltage AC devices with your Raspberry Pi or Arduino.

Controlling AC 13

Switching AC appliances requires an understanding of the dangers and special precautions that must be used when dealing with high voltages. In this chapter, you will learn how to safely control AC devices using electromechanical and solid-state relays, as well as techniques such as zero-crossing switching.

High Voltage Can Kill

Every year, domestic AC electricity kills hundreds of people in the United States alone. Many more are badly burned and many house fires are started by faulty wiring. The domestic AC supply is high voltage and capable of supplying large currents.

*When working on the practical section of this chapter, **never** work on a "hot" (live) line. I like to see the plug of whatever I am working on in front of me on my bench, so I know it's not in the wall outlet.*

Always use an RCD-protected AC outlet while you are working on a project, and never leave exposed wiring or PCBs on any project that controls AC without enclosing it in an insulated box and fixing it down.

Also, do not be tempted to use breadboard with AC designs. It is not designed to cope with the voltages and currents involved.

In short, unless you have had training in working with domestic AC, use a ready-made module like the PowerSwitch tail.

AC Switching in Theory

In this section, you will learn about the theory of AC switching and various circuit designs and components suited to this area of electronics. There is a separate section later in this chapter on switching AC in practice ("AC Switching in Practice" on page 260).

What Is Alternating Current?

Whereas direct current (DC) always has current flowing in one direction, alternating current (AC) just can't keep still. The direction of current flow reverses 120 times per second in some parts of the world (including the United States) and 100 times per second in other parts of the world. This reversing of the current flow is achieved by a reversal of the voltage across the load (Figure 13-1). There are two reversals of the flow of current in each full cycle and so the frequency of the AC is either 60 or 50Hz (cycles per second). Hertz is the unit of frequency and 1Hz is one cycle per second.

In the United States and a few other countries, the AC supply is generally 120V, and in much of the world, AC outlets are more commonly at the far more lethal 220V.

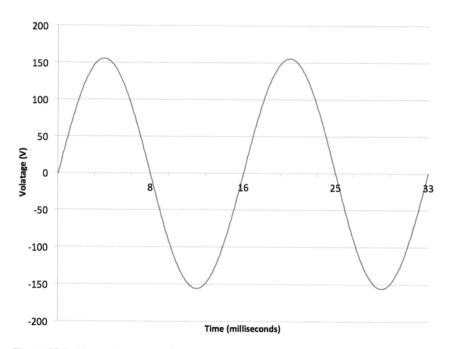

Figure 13-1 *Alternating current (and voltage)*

As you can see from Figure 13-1, the peak positive and negative voltages are actually considerably higher than 120V. The figure 120V AC is actually a kind of average voltage called the RMS voltage. This is the equivalent DC voltage that would be able to supply the same amount of power to a load. So, 120V DC would make an old-fashioned filament light bulb shine at the same brightness as powering it from 120V AC.

Relays

You first worked with relays back in "Controlling DC Motors with a Relay" on page 102 when you used a relay to switch a DC motor. The coil of a relay is isolated from the switching part, which is essential for safe AC switching. Most relays will have their switching capabilities marked on the package. For example, a typical "sugar cube" relay might say that it can switch 10A at 250V AC and 10A at 24V DC.

Figure 13-2 shows how a relay might be used to switch an AC load. Remember that the relay coil requires a bit too much current to be driven directly from the digital output of an Arduino or Raspberry Pi.

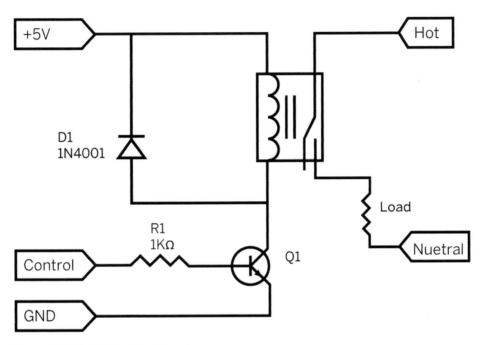

Figure 13-2 *Switching AC with a relay*

The Control, 5V, and GND lines would be connected to your Arduino or Raspberry Pi. When Control is high, the transistor Q1 will turn on, energizing the relay coil that closes the relay contact connecting the AC Hot/Live line to one end of the load (device that you want to switch on and off). The other end of the load is connected to the Neutral AC line.

Optoisolator

Relays are an old technology and solid-state relays (SSRs), which are discussed in "Solid State Relays (SSRs)" on page 262, are replacing them in many applications that switch AC. One key component of an SSR is an optoisolator. An optoisolator acheives the essential goal of separating the low-voltage switching side of the project from the dangerous high-voltage AC side.

Figure 13-3 shows a transistor-based optoisolator.

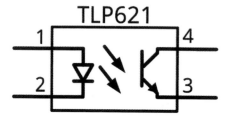

Figure 13-3 *An optoisolator*

An optoisolator combines an LED and a light-sensing element (usually a phototransistor) into a single plastic package.

The key thing is that there is no electrical connection between the LED and the phototransistor, but only an optical link. When the LED is emitting light, the phototransistor will conduct. The phototransistor is low power and will need more electronics before it can control anything AC.

The devices are pretty sensitive and so you can drive the LED side from an Arduino or even Raspberry Pi GPIO pin using a 1kΩ resistor (limiting the current to a couple of milliamps).

Zero-Crossing Optoisolators and Triacs

Optoisolators that are designed to switch AC have a few special features. First, at the light-sensing end, they do not have a normal bipolar phototransistor, but instead use a device called a photo-TRIAC (TRIode for AC). Figure 13-4 shows the internal design of such a device (for example, the MOC3031). You can download the datasheet (*http://www.farnell.com/datasheets/1639837.pdf*) for this device in PDF format.

A TRIAC is a specialized type of transistor designed to switch current flowing in both directions, which is just what you need to control AC.

One feature of a TRIAC is that once conducting, it latches on. It will stay on until the current flowing through it reduces to close to nothing. This makes it pretty useless for controlling DC, but because AC swaps polarity 120 times a second, the TRIAC will have the opportunity to turn off 120 times per second.

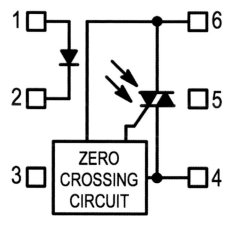

Figure 13-4 *A zero-crossing optoisolator*

An advantage of this latching behavior is that because the TRIAC will only turn off when the current through it (and hence the voltage across it) is low, this reduces the large switching current that would otherwise occur. Switching the current in this way also reduces electrical interference. This gentle switching action is enhanced by the use of the zero-crossing circuit included in some optoisolators. This delays the turning on of the TRIAC until the voltage crosses zero, ensuring that the turn on and the turn off are both smooth.

A typical circuit using a zero-crossing TRAIC to control a higher-power TRIAC is shown in Figure 13-5.

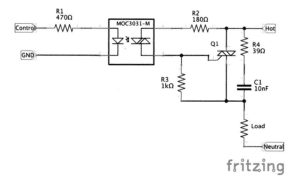

Figure 13-5 *Switching AC with a zero-crossing optoisolator*

The TRIAC that is built into an optoisolator like the MOC3031 is low current and only intended to be used to control a more powerful TRIAC that actually switches the AC.

Controlling the load from an Arduino or Raspberry Pi becomes just a matter of supplying a milliamp or two to the LED inside the optoisolator.

R2 and R3 limit the current flowing through the low-power photo-TRIAC in the optoisolator and R4 and C1 are there to "snub" any voltage transients that arise despite the gentle switching.

AC Switching in Practice

Don't try to build the circuits described in this chapter on breadboard—it's not safe. If you want your Arduino or Raspberry Pi to switch AC loads, follow the guidance in this section.

Relay Modules

Back in "Relay Modules" on page 105 we explored ready-made relay modules. These have the big advantage when dealing with AC that you can connect the AC device that you want to switch using the screw terminals. The relay will naturally isolate the low-voltage side of the project from the dangerous high-voltage part of the design.

Relays do contain metal parts and could potentially fail in such a way that the hot/live side of the relay becomes connected to the relay coils. This could happen if, say, the relay was bashed hard or accidentally crushed. For this reason, relays often have the additional level of safety of using an optoisolator.

Using a Relay Module Safely

*Always remember that a relay module like the one shown in Figure 13-6 will have bare metal conductors and solder joints on both the top and bottom of the board that will be hot/live. Touch these and it could be the last thing you do, so **always** have a plastic enclosure around the relay module and any other parts of the project so that your (or anyone else's) fingers can't inadvertently touch something. The parts inside the box should also be securely anchored to prevent anything from moving around.*

Note the use of strain relief grommets that stop the leads pulling out and possibly becoming loose from the relay module's screw terminal and shorting to something.

Don't work on the relay module or connect wires to the screw terminals when it's connected to the AC, and fit the lid onto the box as soon as you have wired it up.

Figure 13-6 *A relay module in a plastic enclosure*

Also watch out for low-cost relay modules that are sold as being fit for AC use, as many of these are unsafe.

Do not be fooled into thinking that if it says 10A at 25V on the relay itself then the relay module as a whole is good for that. The small screw terminals on many low-cost modules are only rated at 2A. Some of these low-cost boards also have the relay contact solder pads very close (sometimes only a mm or two) to the low-voltage side of the circuit. This is dangerous at high AC voltages, and such relays should only be used for low-voltage DC at modest currents. The best relay modules have an optoisolator and a slot cut in the PCB around the COM relay contact for maximum isolation.

The safest way to switch AC is to use a ready-made and enclosed module like the PowerSwitch Tail (see "The PowerSwitch Tail" on page 263).

You should also check that your relay module is active-low or active-high. If you have an active-low relay module, then it will activate the relay when the digital output is LOW. This means that as soon as you set the pin to be a digital output, you also need to set the output to be HIGH on the next line, otherwise the relay could briefly turn on each time the Arduino resets:

```
pinMode(relayPin, OUTPUT);
digitalWrite(relayPin, HIGH);
```

When using an active-low relay module with a Raspberry Pi, you can make use of the optional parameter initial to set the output HIGH:

```
GPIO.setup(relay_pin, GPIO.OUT, initial=True)
```

It is common for relay modules that have an optoisolator to also have the option of removing a jumper link to allow the positive supply to the relay coil to be provided independently of the positive side of the optoisolator's LED. This provides an additional level of isolation, but does mean that a separate power supply is needed.

Solid State Relays (SSRs)

Figure 13-7 shows an SSR module. This is a sealed and enclosed unit that contains a circuit that is probably very similar to the one shown in Figure 13-5.

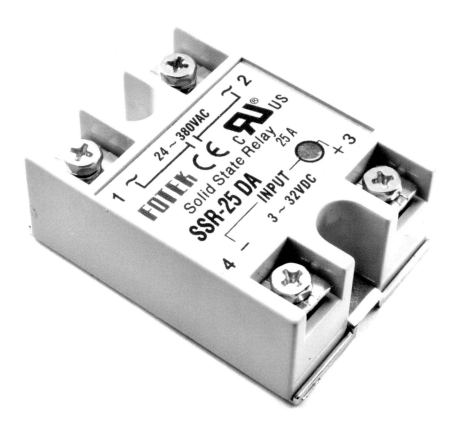

Figure 13-7 *An AC SSR*

These devices are readily available and make AC switching very easy. You still have the problem that there are exposed metal parts that will be hot/live, and so the whole device needs to be enclosed in an insulated box.

The device's low-voltage side can be connected directly to a Raspberry Pi or Arduino, as they include a suitable series resistor for the LED.

The PowerSwitch Tail

The PowerSwitch Tail (Figure 13-8) is an SSR that has an AC plug on one end and an AC outlet on the other.

Figure 13-8 *A PowerSwitch Tail*

There are screw terminals that connect to the LED side of the optoisolator (series resistor built-in) and a small red LED that also lights when the SSR is activated. You will use one of these very handy devices in "Project: Raspberry Pi Timer Switch" on page 263.

Project: Raspberry Pi Timer Switch

This project uses a Raspberry Pi and a PowerSwitch Tail to control the power to a small electrical appliance. In Chapter 16 this very basic project will be improved to add a web interface, so that you can use a browser to switch the appliance on and off (see "Project: A Raspberry Pi Web Switch" on page 301).

This project is very easy to make; the only tool you will need is a screwdriver to unscrew and then tighten the screw terminals.

Parts List

In addition to your Raspberry Pi, you will need the following parts to build this project:

Part	Sources
PowerSwitch Tail	Adafruit: 268
Female-to-male jumper wires	Adafruit: 826
Table lamp or other small appliance	

Construction

Figure 13-9 shows the wiring diagram for the project.

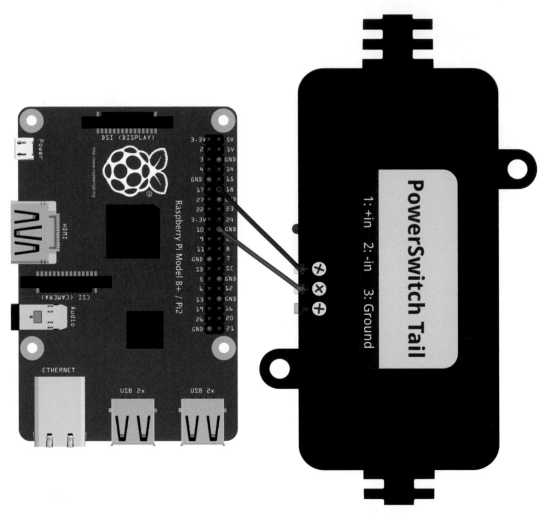

Figure 13-9 *Wiring diagram for the timer switch project*

If you look closely at the PowerSwitch Tail's label, it says that the input is 3-12V DC at 3-30mA. The current required for the input will vary according to the voltage, so the lower current end of the range corresponds to the 3V input. Actually, at 3.3V, the PowerSwitch Tail does draw about 6mA, which is fine for a single Raspberry Pi GPIO pin.

You don't need to attach anything AC to the PowerSwitch Tail while you are testing because its status LED will light when the SSR is switched on.

Software

You can find the software for this project in */python/projects/ac_timer_switch.py* (for information on installing the Python code for the book, see "The Book Code" on page 34 in Chapter 3):

```python
import RPi.GPIO as GPIO
import time

GPIO.setmode(GPIO.BCM)

control_pin = 18

GPIO.setup(control_pin, GPIO.OUT)

try:
    while True:          ❶
        duration_str = input("On time in minutes: ") # ❷
        duration = int(duration_str) * 60   ❸

        GPIO.output(control_pin, True)   ❹
        time.sleep(duration)
        GPIO.output(control_pin, False)   ❺

finally:
    print("Cleaning up")
    GPIO.cleanup()
```

The program is also very simple; it starts with the usual imports and constant definitions.

❶   While `True` is a way of making the loop continue forever, as the condition `True` is never `False`. This is the only reason a while loop ever finishes (unless you press Ctrl-C).

❷   The main loop prompts you to enter a number of minutes that you want the light to be on for.

❸   Convert the string value of `duration` as a string into an integer number of minutes using `int` and then multiply by 60 to convert it into seconds.

❹   The GPIO pin 18 is set to high (`True`) to switch on the PowerSwitch Tail, turning on whatever is plugged into its AC outlet.

❺   After the appropriate delay, the GPIO pin is set `LOW` to turn off the SSR and the loop starts again, prompting you for a new "on time."

Using the Project

The PowerSwitch Tail can switch up to 15A, so you can plug most things into it apart from very high-power devices like electric kettles or hair dryers. A small table lamp might be a good starting point.

Run the program and when it prompts for an "on time" enter 1 for 1 minute and press Enter. Whatever is plugged into the PowerSwitch Tail should turn on and the little status LED on the PowerSwitch Tail itself should also light. At the end of the minute, the Power-Switch Tail should switch off.

Summary

This chapter highlighted some of the dangers of using high-voltage AC, but also showed that it's pretty easy to switch things on and off with the right hardware.

In the next chapter, you will learn about using displays with an Arduino or Raspberry Pi.

Displays

14.

In addition to attaching a monitor to your Raspberry Pi, there are many different output devices that can display text or numbers, graphics, or indeed just let you control a whole load of LEDs in one go. In fact, there are far too many to cover every type of display, so in this chapter, I'll stick to the most practical and fun displays, using a variety of interfacing techniques.

LED Strips

The RGB LEDs described in Chapter 6 are just made up of three LED light emitters in one LED package. Another type of LED called *addressable LEDs* add a chip to the LED package. This chip provides PWM control to the three colors of the LED just like we did with an Arduino or Raspberry Pi in Chapter 6, except that this driver chip is built into the LED.

These addressable LEDs are designed to be controlled in large numbers using a single microcontroller or computer such as an Arduino or Raspberry Pi. Adafruit sells addressable LEDs that they have named NeoPixels (the term NeoPixels is often used to describe non-Adafruit addressable LEDs, especially on eBay). The most used addressable LEDs are of the type WS2812. They use a serial data standard that allows long chains of these LEDs to be connected together to make big displays. As well as RGB-addressable LEDs, they are also available as single colors.

Sometimes they are arranged in a matrix but you can also buy them on a reel of tape and can actually cut the LED strip (Figure 14-1) to suit the number of LEDs you need for your project.

Figure 14-1 *NeoPixel LED strip*

If you look closely at Figure 14-1, you can see that the strip is in segments like a tape worm. The three solder pads on each side of the LED carry the power GND and 5V and also the serial data D0 that is "daisy-chained" from one LED to the next. In other words, the output of one LED is connected to the input of the next. The inviting line down the middle of the solder pads is where you cut the LED strip if you want to shorten it.

Notice also the arrow on the strip that indicates the direction that the serial data flows. So, you always connect on the left to control the LEDs to the right.

Experiment: Controlling an RGB LED Strip Display

After purchasing a meter or two of LED strip from either eBay or Adafruit you'll probably be keen to try it out. It is easy to connect to an Arduino (Figure 14-2), but requires a little more work to connect to a Raspberry Pi.

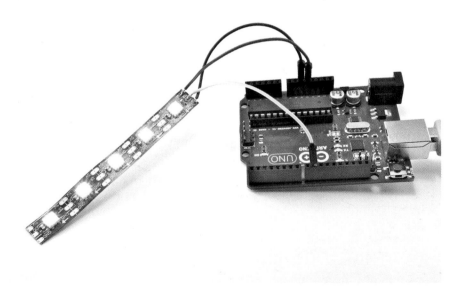

Figure 14-2 *LED strip and Arduino*

Parts List

You need the following parts to carry out this experiment using an Arduino:

Part	Sources
WS2812 Addressable LED strip	eBay, Adafruit: 1376
3 x male-to-male jumper wires	Adafruit: 758

If you are using a Raspberry Pi, you will also need the following items, as the logic input for the NeoPixels is not specified as being compatible with 3V logic. Having said that, you may like to try it without the logic-level converter, as your LED strip may work just fine with 3V logic before you get the extra bits.

If you want to try this on a Raspberry Pi, you might also need the following parts:

Name	Description	Source
	400-point solderless breadboard	Adafruit: 64
R1, R2	2 x 470Ω 1/4 W resistor	Mouser: 291-470-RC
Q1	2N7000 MOSFET transistor	Mouser: 512-2N7000
	Female-to-male jumper wires	Adafruit: 826

This project departs from our standard list of transistors to use a low-power MOSFET transistor for the logic-level conversion, as it responds much better to the high-frequency serial data than a 2N3904, for example. If you prefer, you could use an FQP30N06L, but this high-power device would be serious overkill for the project.

Arduino Connections

At one end of the reel, there will often be a cable with a three-pin plug on the end connected to the GND, 5V, and D0 connections of the strip. If so, you can use female-to-male jumpers to connect this to a breadboard, or to your Arduino. But, if you have already cut off some of your pixels for a project, you will need to carefully solder some leads to the display.

I sacrificed three male-to-male header leads, chopping off the pins at one end and soldering the leads to the strip of five NeoPixels, as shown in Figure 14-3.

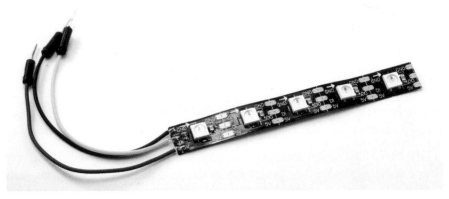

Figure 14-3 *A convenient NeoPixel display*

Addressable LED Power Consumption

When cranked up to maximum brightness and white, the addressable LEDs draw quite a lot of current (about 60mA per LED). While five addressable LEDs could consume 300mA (which is just fine direct from an Arduino or Raspberry Pi), any more than that and you probably want to start considering supplying 5V direct to the LED strip from its own power supply.

The three header pins can then plug directly into the Arduino with the LED strip's data pin D0 connected to D9 on the Arduino, as shown in Figure 14-2.

Arduino Software

The helpful folks at Adafruit have created an Arduino library that makes it easy to control long strings of addressable LEDs. You can download this library from *https://github.com/ adafruit/Adafruit_NeoPixel* and install it into your Arduino environment (see "Installing Arduino Libraries" on page 225 in Chapter 12).

You can find the example Arduino sketch for this experiment in */arduino/experiments/ neopixel* (see "The Book Code" on page 14 in Chapter 2 for guidance on installing the Arduino sketches for the book):

```
#include <Adafruit_NeoPixel.h>   ❶

const int pixelPin = 9;     ❷
const int numPixels = 5;    ❸

Adafruit_NeoPixel pixels = Adafruit_NeoPixel(numPixels, pixelPin, NEO_GRB +
NEO_KHZ800); // ❹

void setup() {
  pixels.begin(); / ❺
}
```

```
void loop() {
  for (int i = 0; i < numPixels; i++) {  ❻
    int red = random(255);
    int green = random(255);
    int blue = random(255);
    pixels.setPixelColor(i, pixels.Color(red, green, blue)); // ❼
    pixels.show();
  }
  delay(100L);
}
```

❶   Import the Adafruit NeoPixel library.

❷   Change this if you want to use a different pin to drive the NeoPixels.

❸   Change this if you have more or LEDs on your strip, but check "Addressable LED Power Consumption" on page 270 before you add too many LEDs.

❹   Initialize the NeoPixel library for your setup.

❺   Start sending the display data.

❻   Assign a random color for each of the pixels, update the display, and then delay for 1/10 second before changing all the LED colors again. Disco time!

❼   The setPixelColor function takes two parameters: the index position of the pixel to set and the color, which is itself made up of three values between 0 and 255 for each of the three color channels.

Raspberry Pi Connections

You might find that your Raspberry Pi will work just fine without level conversion. So try that before taking the breadboard and level converter approach.

So, just take the 5 NeoPixel display that you made to use with an Arduino and use female-to-female jumper wires to make the following connections:

- GND NeoPixel to GND Raspberry Pi
- 5V NeoPixel to 5V Raspberry Pi
- D0 NeoPixel to GPIO 18

Figure 14-4 shows the LED strip wired directly to the Raspberry Pi.

Figure 14-4 *NeoPixels connected directly to a Raspberry Pi*

Jump ahead to "Raspberry Pi Software" on page 273 and try out the program *neopixel_no_level_conv.py*.

If your LED strip didn't light up and make a nice colorful display, then you probably need to do the job properly and use a level converter to raise the 3V control signal up to 5V.

From 3 to 5V

Although you may be lucky, the 3V control signal from a Raspberry Pi is below the minimum of 4V expected by a WS2812 addressable LED when it's expecting HIGH. The arrangement shown in Figure 14-5 shows how you can use a MOSFET transistor to shift the signal level up to 5V.

A side effect of this level shifting is that the output is inverted. That is, if the Raspberry Pi supplies logical LOW (0V) the output to the LED strip will be 5V and when the Raspberry Pi GPIO pin is HIGH (3.3V) the output to the LED strip will be 0V.

Fortunately, this is something that can be easily corrected for in the software.

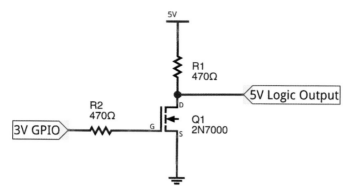

Figure 14-5 *Level conversion 3 to 5V*

Figure 14-6 shows the breadboard layout for the experiment with level conversion.

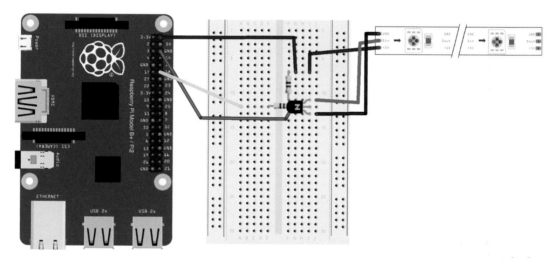

fritzing

Figure 14-6 *Connecting a Raspberry Pi with level conversion*

Raspberry Pi Software

The software used here is based on the Adafruit tutorial (*https://learn.adafruit.com/ neopixels-on-raspberry-pi*). However, the C library used in the Adafruit tutorial is (at the time of writing) not compatible with the Raspberry Pi 2.

Fortunately for Raspberry Pi 2 owners, Richard Hurst has made a version of the software that does work with Raspberry Pi 2. This version also works with earlier versions of the Raspberry Pi. Install the packages and library software that are needed using this command:

```
$ sudo apt-get install build-essential python-dev git scons swig
```

The next step is where the directions diverge from the Adafruit tutorial, because you need to fetch the version of the software that has been fixed for Raspberry Pi 2.

Fetch the modified NeoPixel code from GitHub using this command:

```
$ git clone https://github.com/richardghirst/rpi_ws281x.git
```

Change directory into the software that you just fetched from GitHub and then build the C code using the command:

```
$ scons
```

When the C is compiled, you need to install the Python library that interfaces with the fast C code using:

```
$ cd python
$ sudo python setup.py install
```

There are two almost identical versions of the Python program: one adjusted to work with an inverting level converter and one designed to work with the LED strip connected directly to the Raspberry Pi. So run either *neopixel_no_level_conv.py* or *neopixel.py* (if you have the breadboard circuit) depending on how you have your hardware set up. Both programs can be found in *python/experiments/*.

The following listing shows the version assuming an inverting level converter circuit:

```
import time, random
from neopixel import *        ❶

# LED strip      ❷
LED_COUNT      = 30        # Number of LED pixels.
LED_PIN        = 18        # GPIO pin connected to the pixels (must support PWM!).
LED_FREQ_HZ    = 800000  # LED signal frequency in hertz (usually 800khz)
LED_DMA        = 5         # DMA channel to use for generating signal (try 5)
LED_BRIGHTNESS = 255       # Set to 0 for darkest and 255 for brightest
LED_INVERT     = True      # True to invert the signal (when using NPN transistor level
shift)

# Initialize the display
strip = Adafruit_NeoPixel(LED_COUNT, LED_PIN, LED_FREQ_HZ, LED_DMA, LED_INVERT,
LED_BRIGHTNESS)
strip.begin()

while True:        ❸
    for i in range(strip.numPixels()):
        red = random.randint(0, 255)
        green = random.randint(0, 255)
        blue = random.randint(0, 255)
        strip.setPixelColor(i, Color(red, green, blue))
        strip.show()
    time.sleep(0.1)
```

❶   Import the NeoPixel library.

❷ This set of parameters shouldn't need changing, unless you want to use a different LED_PIN.

❸ The main loop of the program works much like its Arduino counterpart, generating a random color and then assigning it to a pixel.

I2C OLED Displays

Although a Raspberry Pi can be connected to any monitor (large or tiny) that has an HDMI or AV socket, sometimes you just need a couple of lines of text to be displayed. An Arduino Uno does not have video output and so again a small display that can show a bit of graphics or a couple of lines of text is very useful.

Small, organic LED (OLED) displays are cheap, don't use much current, and are really clear and easy to read (Figure 14-7). They are replacing LCD displays in many consumer products. They are available as monochrome and color displays.

Figure 14-7 *An OLED display*

A common type of OLED display can be found as a module that includes the display itself and a driver chip on a circuit board. The driver board generally uses the SSD1306 chip that has an I2C (pronounced "i squared c") interface that requires just two pins for data and two pins for power.

These screens are small and high resolution, so if you're using an Arduino, you can sometimes run a little short on memory, if you're not careful.

Experiment: Using an I2C Display Module with Raspberry Pi

I2C OLED displays can be used with Arduino and Raspberry Pi. In "Project: Adding a Display to the Beverage Cooler Project" on page 279, you can replace the temperature LED with an OLED display that shows both the current and set temperature.

This experiment is just to show you how to use such a display with a Raspberry Pi. The example code displays the time and a little animation (Figure 14-8).

Figure 14-8 *A clock using an OLED display*

Parts List

The OLED display will have four header pins for connection. These can be connected directly to the Raspberry Pi using four female-to-female header wires.

Part	Sources
I2C OLED display 128x64 pixels	eBay
4 x female-to-female jumper wires	Adafruit: 266

Look for a display that is 128x64 pixels that uses an SSD1306 driver chip. Some of these displays use extra pins from the SSD1306 that are not really needed, so for simplicity, look for a display with just four pins: GND, VCC (+V), SDA (Data), and CLK (Clock). The display I used was monochrome, but a color one should work fine too.

If you use one of the Adafruit displays, follow the instructions on the Adafruit site (*https://learn.adafruit.com/096-mini-color-oled/*) to wire up the extra pins.

Connections

The SSD1306 will work at 3V or 5V. Because earlier Raspberry Pi models could only supply quite small currents at 3V, it makes sense to power it from 5V.

Make the following four connections with the header leads:

- GND on the display to GND on the Raspberry Pi
- VCC on the display to 5V on the Raspberry Pi
- SCL on the display to GPIO 3 on the Raspberry Pi
- SDA on the display to GPIO 2 on the Raspberry Pi

Software

If you have not already set up your Raspberry Pi to support I2C, refer back to "Setting Up I2C on Your Raspberry Pi" on page 171.

Adafruit has produced a great Python library for using these displays that works on pretty much any SSD1306 display, whether sold by Adafruit or another vendor. To download the library, you can clone the library's repository directly onto your Raspberry Pi using this command:

```
$ git clone https://github.com/adafruit/Adafruit-SSD1331-OLED-Driver-Library-for-
Arduino.git
```

To install the library, change into the directory that the `clone` command creates and then run the installer script:

```
$ cd Adafruit_Python_SSD1306
$ sudo python setup.py install
```

You can find the example program for this experiment in the *python/experiments/oled.py* file.

Coordinates

The OLED display is a graphical display that you can draw various shapes to and also write text. You, however, need to specify the positions of the things that you want to display on the screen. The Adafruit library uses the coordinate system shown in Figure 14-9.

All pixel positions are specified from the top-left corner of the screen, with the bottom-right pixel having the coordinates x=127, y=63.

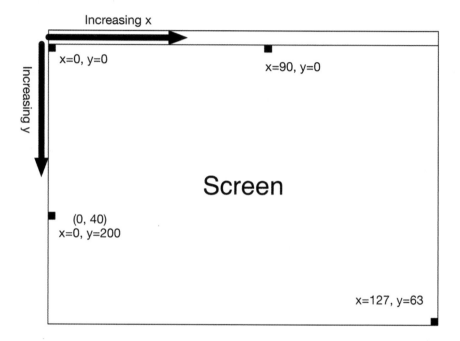

Figure 14-9 *The display coordinate system*

```
from oled.device import ssd1306      ❶
from oled.render import canvas
from PIL import ImageFont      ❷
import time

device = ssd1306(port=1, address=0x3C)      ❸
large_font = ImageFont.truetype('FreeMono.ttf', 24)  # ❹

x = 0
while True:
    with canvas(device) as draw:      ❺
        draw.pieslice((x, 30, x+30, 60), 45, -45, fill=255)   # ❻
        x += 10      ❼
```

```
        if x > 128:
            x = 0
        now = time.localtime()      ❽
        draw.text((0, 0), time.strftime('%H:%M:%S', now), font=large_font, fill=255)
    time.sleep(0.1)
```

❶   There are two imports from the OLED library: one of `ssd1306` for the device itself and one for `canvas`, which is the Python object onto which shapes and text are drawn.

❷   The `PIL` (Python Image Library) is also imported.

❸   Create a variable (`device`) to gain access to the display. The value of `0x3C` used here is the I2C address of the display. This is commonly used in low-cost eBay displays, but you will find that other displays, including the Adafruit displays, use a different address here. Refer to the documentation for your display.

❹   Specify a font to use of height 24 pixels. This will be used when the time is written to the display.

❺   This `with as` construct is used to keep all the text drawing code together.

❻   Draw a Pac-Man shape of size 30 x 30 pixels from an angle of 45º to –45º to give the circle a mouth. The fill color of 255 means white.

❼   Add 10 to `x`. The variable `x` sets the starting position of the `pieslice` drawing and so by increasing it by 10 each time around the loop we can achieve a crude animation.

❽   Get the current time, format it into a string, and then write it to the display.

Experimentation

Run the program using:

```
$ sudo python oled.py
```

If the screen remains blank, the most likely problem is that the I2C address is not correct.

In this example the `piesclice` function was used. This function draws a circle with a slice cut out of it. There are lots of other graphics drawing functions that you can use, so take a look at *http://effbot.org/imagingbook/imagedraw.htm* and try out a few of them to spice up your display.

Project: Adding a Display to the Beverage Cooler Project

As an example of using an OLED display with an Arduino, this project builds on "Project: A Thermostatic Beverage Cooler" on page 246 to replace the green LED that shows when the cooler is at about the right temperature with an OLED display to show both the actual and set temperatures (Figure 14-10).

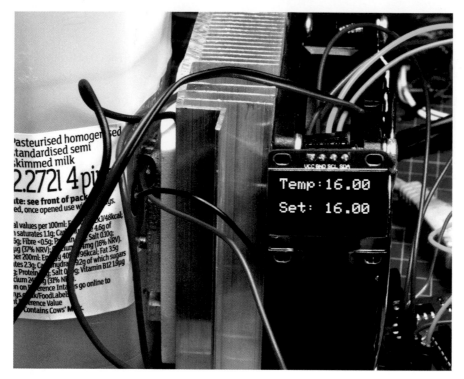

Figure 14-10 *Adding an OLED display to the cooler project*

Parts List

In addition to the parts used for "Project: A Thermostatic Beverage Cooler" on page 246 (less the LED and R3), you will need an OLED display like the one in "Experiment: Using an I2C Display Module with Raspberry Pi" on page 276 and four extra female-to-male jumper wires.

Connections

If you have not already done so, you will need to build the project described in "Project: A Thermostatic Beverage Cooler" on page 246. You don't need to add the LED and R3.

The OLED display can connect directly to the Arduino, which reduces the clutter on the breadboard. Figure 14-11 shows just the new part of the project with the OLED display connected to the Arduino.

Be careful to check the pinout of your OLED module, some swap over the 5V and GND pins.

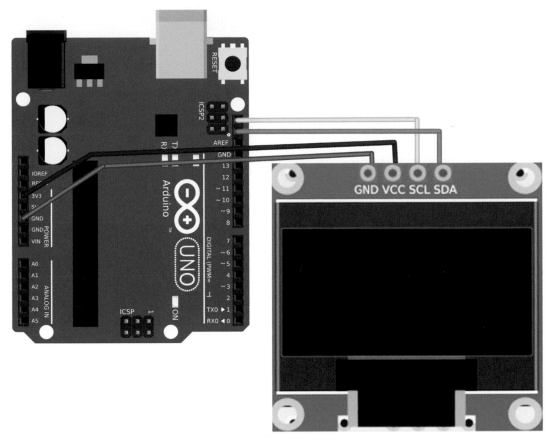

Figure 14-11 *Connecting the OLED display to the Arduino*

Software

As you would expect, there is also an Arduino library (*https://github.com/adafruit/Adafruit_SSD1306*) for the OLED display, so download and install it into your Arduino IDE (see "Installing Arduino Libraries" on page 225 in Chapter 12).

The sketch for this version of the cooler is for the most part exactly the same as without the display. You can find it in *pr_thermostatic_cooler_display*. The changes are the inclusion of the Adafruit GFX and SSD1306 libraries. The SSD1306 library is for the device itself and GFX provides useful functions for displaying text and drawing shapes and so on:

```
#include <Adafruit_GFX.h>
#include <Adafruit_SSD1306.h>
```

The variable `display` is defined to provide a reference to the library. The parameter 4 here is actually a pin number for the Enable pin of the display if it has one. Our display does not have an Enable pin so 4 is chosen as the pin is not used by the Arduino for anything else:

```
Adafruit_SSD1306 display(4);
```

setup() includes the following line that initializes the display:

```
display.begin(SSD1306_SWITCHCAPVCC, 0x3c);
```

The checkTemperature function now includes a call to the new function updateDisplay to refresh the contents of the display with the current temperature every second:

```
void updateDisplay() {
  display.clearDisplay();
  display.setTextSize(2);
  display.setTextColor(WHITE);
  display.setCursor(0,0);
  display.print("Temp:");
  display.println(measuredTemp);
  display.print("Set: ");
  display.println(setTemp);
  display.display();
}
```

The function readSetTempFromPot has been modified slightly so that now, if the set temperature changes then updateDisplay is called immediately so that when you turn the pot, the set temperature updates without having to wait until the next second:

```
double readSetTempFromPot() {
  static double oldTemp = 0;
  int raw = analogRead(potPin);
  double temp = map(raw, 0, 1023, minTemp, maxTemp);
  if (oldTemp != temp) {
    updateDisplay();
    oldTemp = temp;
  }
  return temp;
}
```

The keyword static before the definition of oldTemp means that oldTemp will keep its value between successive calls to readSetTempFromPot. The variables temp and oldTemp are both of type double (double-precision float) because that is the type that the DS18B20 library uses.

Summary

There are many different types of display, and this chapter has just touched on a few of the most useful.

In the next chapter, you will learn about using Arduino and the Raspberry Pi to create sound.

Sound | 15

Having looked at movement, light, heat, and displays, it's time to turn our attention to making sound.

High-quality audio is easy on the Raspberry Pi because powered speakers can be attached to the audio jack, but things are a little more difficult with the Arduino.

Experiment: Unamplified Speaker and Arduino

Ultimately, if you want to make any kind of complex sound, you are going to need some kind of loudspeaker. Loudspeakers have been around for almost 100 years and work a little like a solenoid (see "Solenoids" on page 122) that pushes a rigid cone at a high enough frequency to make sound waves.

Loudspeakers are generally marked with a value in ohms. This value is like their resistance but is actually their *impedance*. Impedance, as the word suggests, is like resistance but applies to things that are not pure resistors and the coil of wire in the speaker (like any coil) does not act quite like a resistor. If you are interested in such things, read up on *inductance*.

Common values for a loudspeaker are 4Ω or 8Ω. If you were to connect an 8Ω speaker to the 5V output of an Arduino, you could reasonably expect a current of I = V / R = 5 / 8 = 625mA. That's a lot more than the 40mA recommended for an Arduino output pin. It looks like we need a resistor!

In this experiment, you will connect up a speaker to an Arduino via a resistor and then use the Serial Monitor to instruct the Arduino to produce sound of a particular frequency.

Audio Frequencies

The frequency of a sound wave is what in musical terms is called the *pitch* and it is the number of sound waves arriving at your ears per second. I like to think of soundwaves like ripples in a pond. So a high-frequency sound of, say, 10kHz (kilohertz) will have 10,000 sound waves per second and a low frequency (say, 100Hz) will have just 100 waves per second. The limits of human hearing are often taken as being between 20Hz and 20kHz, but the upper limit decreases as you get older. Sounds over 20kHz are generally referred to as ultrasound.

Other animal species have different ranges of frequencies that they can hear. For example, cats can hear frequencies with a top range of 55 to 79kHz and of course bats famously use ultrasonic echo location.

In music, the lowest C on a standard piano has a frequency of 32.7Hz and the highest is 4.186kHz. An octave is doubling of the frequency. So if you take two adjacent C notes on a piano keyboard, the second is double the frequency of the first.

Parts List

You really don't need much for this experiment, just a speaker and resistor, although some breadboard and jumper wires make it easier to connect things up:

Part	Sources
Small 8Ω speaker	Adafruit: 1891
270Ω 1/4 W resistor	Mouser: 291-270-RC
400-point solderless breadboard	Adafruit: 64
Male-to-male jumper wires	Adafruit: 758

The speaker that I used was scavenged from an old radio receiver and had a connector on the end into which a male-to-male jumper wire could be inserted. You may find a speaker with wires attached that will push into the breadboard or Arduino sockets, or you might have to solder some leads on with wires that are thin enough to fit into breadboard holes.

Breadboard Layout

Figure 15-1 shows the setup for the experiment.

Figure 15-1 *Arduino and speaker*

One connection to the speaker is connected to an Arduino GND socket and the other is connected via the resistor to pin D11.

Arduino Software

Here's the sketch for this project, which you will find in */arduino/experiments/ ex_speaker* (for information on installing the Arduino sketches for this book, see "The Book Code" on page 14 in Chapter 2):

```
const int soundPin = 11;

void setup() {
  pinMode(soundPin, OUTPUT);
  Serial.begin(9600);
  Serial.println("Enter frequency");
}

void loop() {
  if (Serial.available()) {
    int f = Serial.parseInt();
    tone(soundPin, f);        ❶
    delay(2000);
    noTone(soundPin);         ❷
  }
}
```

The part of this code that might look unfamiliar to you is inside loop().

❶   tone sets one of the output pins of the Arduino to play a tone at the frequency specified (in this case, the frequency that you typed into the Serial Monitor).

❷ After a two-second delay, the noTone command cancels to tone, restoring a restful silence.

Arduino Experimentation

Upload the program and then open the Serial Monitor. Try entering 1000. This should produce a not particularly pleasant sound, at a reasonable volume, but not loud enough to be heard in a noisy room.

Experiment entering different frequencies and see how the note changes.

Although it might be tempting to try and test the frequency range of your hearing with this setup, sadly this won't work because the speaker is quite likely to have a restricted range itself. At frequencies over about 10kHz, the level of sound it produces will fall off very sharply. Similarly, a small speaker will not normally be able to produce sound below about 100Hz.

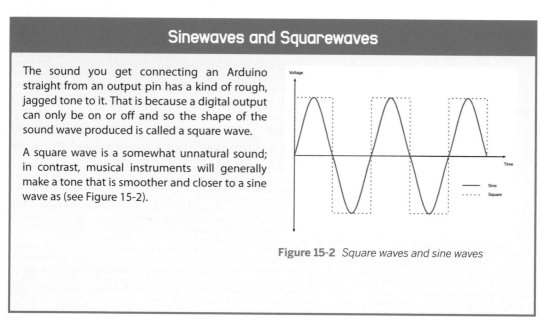

Sinewaves and Squarewaves

The sound you get connecting an Arduino straight from an output pin has a kind of rough, jagged tone to it. That is because a digital output can only be on or off and so the shape of the sound wave produced is called a square wave.

A square wave is a somewhat unnatural sound; in contrast, musical instruments will generally make a tone that is smoother and closer to a sine wave as (see Figure 15-2).

Figure 15-2 *Square waves and sine waves*

Amplifiers

If you need your sound to be louder, then you need to supply more power to your speaker. In other words, you need to amplify the signal so that more power reaches the loudspeaker.

In the kind of setup you had in "Experiment: Unamplified Speaker and Arduino" on page 283, you could get a lot more power into the speaker using a single transistor as if you were switching a relay or motor on and off. Your ugly noise will still be an ugly noise, but a much louder ugly noise.

If you need to generate a nicer waveform (say for music or speech), then the on/off approach is going to sound pretty awful and you would need to look at using a proper audio amplifier.

While you could make your amplifier from scratch, it's a lot easier to use a ready-made module or even just a set of powered speakers intended to sit next to your PC. Using powered (amplified) speakers is especially attractive when it comes to the Raspberry Pi, as the aux lead can just plug into the Raspberry Pi's audio socket.

You will use this approach later in this chapter when you give Pepe the Puppet (from "Project: Pepe, the Dancing Raspberry Pi Puppet" on page 162) a voice in "Project: Pepe the Puppet Gets a Voice" on page 293.

Experiment: Playing Sound Files on an Arduino

You can actually play WAV sound files on an Arduino using the hardware from "Experiment: Unamplified Speaker and Arduino" on page 283 and an Arduino library called PCM (pulse code modulation). This uses a technique a bit like PWM to generate an approximation to the sound. The Arduino only has enough flash memory for about 4 seconds of recording. If you want to play longer sound clips than this, you will need to add an SD card reader to the Arduino and use an approach like the one described on the Arduino website (*https://www.arduino.cc/en/Tutorial/SimpleAudioPlayer*).

You can record the sound onto your computer using the software package Audacity and then run a utility program to convert the sound file into a series of numbers that represent the sound and can be pasted into an Arduino sketch and then played back.

The original article that describes this approach is at High-Low Tech (*http://highlow tech.org/?p=1963*). This experiment differs a little in that it uses the free Audacity software package to record an audio clip.

Parts List

The hardware for this experiment is just the same as "Experiment: Unamplified Speaker and Arduino" on page 283. However, you will need to install the following software on your computer to be able to record and process an audio clip:

- Audacity (*http://audacityteam.org/*)
- The Audio Encoder utility (look for the link on *http://highlowtech.org/?p=1963* for your operating system)

Creating the Sound Data

If you do not want to record your own sound clip, you can jump ahead to "Arduino Experimentation" on page 290 and run the *ex_wav_arduino* sketch that has a short message encoded in it.

To create the sound file, you first need to start Audacity. Before making your recording, set the recording mode to Mono and the Project Rate to 8000Hz. These options are highlighted in Figure 15-3.

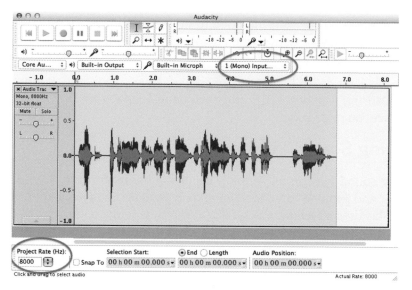

Figure 15-3 *Recording a sound clip*

Click the red Record button to start recording and record your message. Note that this cannot be longer than about 4 seconds. Once you have recorded the message, you will see the sound wave in Audacity. You can select any silent area at the start or end and delete it, to just leave the part of the message that you want.

The next step is to export the sound file. Again this requires certain options to be set. From the File menu, select Export. Then in the Format dropdown, select "Other uncompressed files," and click Options, and select WAV (Microsoft) and Unsigned 8 bit PCM (Figure 15-4). Select a filename and continue past the screen that prompts you for details of the artist.

The file that you have just generated is binary data. This needs to be converted into a list of text numbers, each separated by commas, that can be pasted into your sketch. To do this, run the Audio Encoder utility that you downloaded from *highlowtech.org*. This will prompt you to select the file you want to convert, so select the file that you just exported from Audacity.

After a few moments, you will be shown a dialog window confirming that all the data is in your clipboard.

Open up the sketch */arduino/experiments/wav_arduino*. You are going to replace the whole of the line that starts with 125, 119, 115 with the data in the clipboard. This is a very long line, so the best way to select it is to put the cursor at the start of the line, and then hold the Shift key down while you cursor down and then left one place. Use the Paste option to replace the selected text with the data in your clipboard.

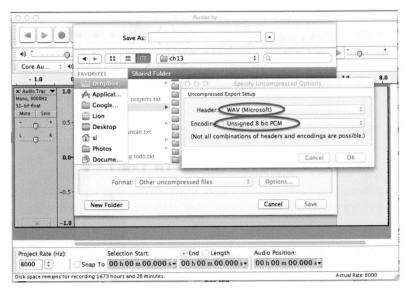

Figure 15-4 *Setting export options*

If you plotted each of those numbers on a chart, the shape you would see would be the same as you saw in Audacity when you were recording a sound clip.

Arduino Code

Before you can compile and run the sketch, you will need to install the PCM library. Download the ZIP archive for the file from GitHub (*https://github.com/damellis/PCM/zipball/master*); unzip it, rename the folder to just *PCM*, and move it into your Arduino libraries folder, as described in "Installing Arduino Libraries" on page 225 in Chapter 12.

The Arduino sketch (if you ignore the sound data) is tiny:

```
#include <PCM.h>

const unsigned char sample[] PROGMEM = {   // ❶
125, 119, 115, 115, 112, 116, 114, 113, 124, 126, 136, 145, 139,
};

void setup() {
  startPlayback(sample, sizeof(sample));  // ❷
}

void loop() { // ❸
}
```

However, that array of data is a very long line!

❶ The data is held in an array of type char, which contains 8-bit unsigned numbers. The PROGMEM command ensures that the data is stored in the flash memory of the Arduino (there should be about 32kB available).

❷ The PCM library plays the sample. startPlayback is passed the array of data to be played and the size of that data in bytes.

❸ The sound clip is played once each time the Arduino resets, so the loop() function is empty.

Arduino Experimentation

Install the sketch onto your Arduino and as soon as it installs, the sound clip will play!

As you upload the sketch, you will see a message at the bottom of the Arduino IDE that tells you how much of the Arduino's flash memory was used. It will say something like "Binary sketch size: 11,596 bytes (of a 32,256 byte maximum)." If the sound file gets too big, you will get an error message.

Connecting an Arduino to an Amplifier

The previous experiment works surprisingly well, given that it is only running on a humble Arduino.

The audio signal coming from the Arduino may be fed through a resistor to keep the current low, but the Arduino pin is operating at 5V, which is too high to act as the input to a typical audio amplifier. So to connect an Arduino to a set of powered speakers to make the sound louder, ironically we first need to reduce its output voltage.

A very convenient way of doing that is to use a pair of resistors as a voltage divider.

Voltage Divider

A voltage divider is a way of using two resistors to reduce a voltage. If that voltage is a varying one, say from an audio signal, then the potential divider scales the voltage by a fixed proportion.

For example, Figure 15-5 shows the voltage divider used to reduce a 5V signal from an Arduino to a much more suitable half volt (more or less) as an input to an audio amplifier.

The voltage where the two resistors connect (Vout) is calculated by the formula:

$$V_{out} = \frac{R_2}{R_1 + R_2} \cdot V_{in}$$

In this case, when the digital output is HIGH, Vin=5V so Vout = 5 * 1/(1+10)=0.45V.

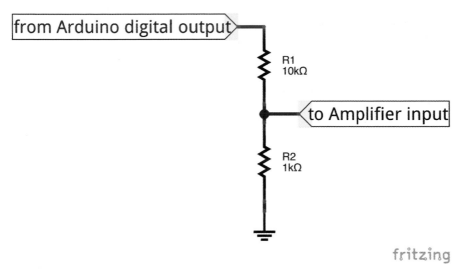

Figure 15-5 *A voltage divider*

You can adapt the breadboard for "Experiment: Unamplified Speaker and Arduino" on page 283, placing one resistor above the other as shown in Figure 15-6, with the top of the 10kΩ resistor connected to D11 and the bottom of the bottom resistor connected to GND. You then just need a way of connecting GND and the breadboard row where the resistors meet to the amplifier.

Figure 15-6 *Attaching an Arduino to an aux lead*

One way to make this connection is to sacrifice an aux lead, cutting it in half and stripping the wires inside the lead to make the connection. Generally you will find three wires, as most leads are stereo. There will be one ground wire, and separate wires for the left and right audio channels. The left and right channels are often red and white.

The only lead that you really need to identify is the ground lead, because the left and right leads are best connected together so that the mono signal from the Arduino is heard from both speakers. You can identify the ground lead using a multimeter set to its continuity (or buzzer) mode (Figure 15-7).

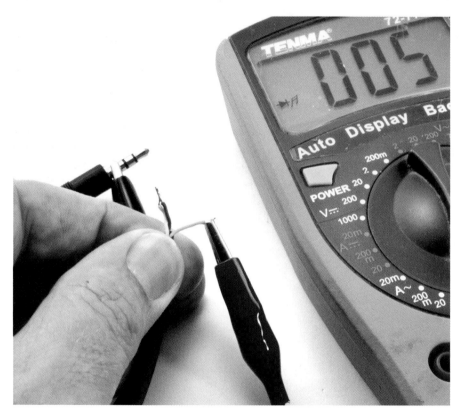

Figure 15-7 *Testing the aux lead*

Clip or touch one of the multimeter leads to the connection of the plug furthest from the plug tip. Then try each of the three leads in turn until the multimeter beeps (or otherwise indicates that there is a connection). That lead is the ground lead and can push into the ground row of the breadboard. The other two leads can be twisted together and go into the output row of the breadboard where the two resistors meet.

Try re-running one of the Arduino experiments earlier in this chapter and you should have a much louder and clearer result.

Playing Sound Files on Raspberry Pi

The Raspberry Pi is a fully fledged computer with an audio output jack. So, playing a sound file on a Raspberry Pi is a matter of finding the right software package to play a sound file.

There are various methods for doing this, but a StackExchange discussion (*http://raspber rypi.stackexchange.com/questions/7088/playing-audio-files-with-python*) covers pretty much all of them.

The method I use here is to use the pygame Python library that is already installed on the Raspberry Pi.

WAV files, while a lot bigger in file size than MP3s, have the great advantage that they are very easy for a Raspberry Pi to decode and so don't slow down your Pi much. Whereas "Arduino Experimentation" on page 290 was quite fussy about the WAV files that it will play, the Raspberry Pi is not limited by memory or processor speed when it come to playing a sound file, so almost any WAV file should play OK.

You can try this out from the Python command line using the sound file that you will use in "Project: Pepe the Puppet Gets a Voice" on page 293. On your Raspberry Pi command line, change directory to the file downloads for the book and within that to the *python/ projects/puppet_voice* folder. In there you will find a file called *pepe_1.wav*. To play this file, connect powered speakers or headphones to the audio socket on your Raspberry Pi and then start the Python console using the command python:

```
>>> from pygame import mixer
>>> mixer.init()
>>> mixer.music.load("pepe_1.wav")
>>> mixer.music.play()
```

You should hear a short message from Pepe.

Project: Pepe the Puppet Gets a Voice

Now that you can make your Raspberry Pi play sound files, you can combine this with "Project: Pepe, the Dancing Raspberry Pi Puppet" on page 162 and a PIR sensor to make Pepe dance and talk whenever someone comes close to him (Figure 15-8).

Figure 15-8 *Pepe with movement detection and voice*

Parts List

To make this project, you will need all the parts from "Project: Pepe, the Dancing Raspberry Pi Puppet" on page 162 and the following extra components:

Part	Sources
Passive infrared (PIR) sensor module	eBay, Adafruit: 189
Female-to-male jumper wires	Adafruit: 826
Powered speakers	
400-point solderless breadboard	Adafruit: 64

PIR sensors are used to detect movement in intruder alarms. This low-cost module is ideal for triggering Pepe when someone comes near. Adding the PIR sensor makes it worthwhile to add a breadboard to hold the servo driver board and wires to the PIR sensor.

Breadboard Layout

In this project, the breadboard provides a firm anchor for the wires to the PIR module and for the servo controller board that will plug directly into the breadboard. Most breadboards have a self-adhesive pad on the underside that you can use to stick it onto the servo chassis to make things a bit more permanent. Figure 15-9 illustrates the breadboard layout for the project, and Figure 15-10 shows the actual wiring of the project.

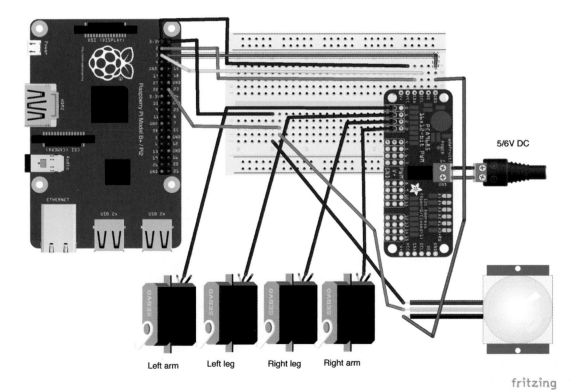

Figure 15-9 *The breadboard layout for the talking puppet*

Figure 15-10 *The talking puppet*

The PIR sensor has a 3V digital output, but requires a 5V power supply. This makes it ideal for use with a Raspberry Pi.

PIR Sensors

Passive Infra Red (PIR) sensors detect movement of anything that emits heat (such as people). It sets a digital output on the PIR high for a second or two every time it registers a change in the level, or in some devices, the pattern of infrared falling on its sensor.

Wiring it up to an Arduino or Raspberry Pi is just a matter of supplying it with power and then connecting its digital output to a digital input on the Raspberry Pi or Arduino.

Software

The software for this project is based on "Project: Pepe, the Dancing Raspberry Pi Puppet" on page 162, so you may want to look at that description of the code too.

You can find all the files for this project in the directory *python/projects/puppet_voice*. As well as the the Adafruit servo code and the program itself (*puppet_voice.py*), you will also find a sound file here called *pepe_1.wav*. This is the sound file that will be played when Pepe is triggered by movement:

```
from Adafruit_PWM_Servo_Driver import PWM
import RPi.GPIO as GPIO
from pygame import mixer
```

```
import time

PIR_PIN = 23      ❶
GPIO.setmode(GPIO.BCM)
GPIO.setup(PIR_PIN, GPIO.IN)

pwm = PWM(0x40)
mixer.init()      ❷
mixer.music.load("pepe_1.wav")

servoMin = 150  # Min pulse length out of 4096
servoMax = 600  # Max pulse length out of 4096

dance = [
  #lh  lf  rf  rh
  [130, 20, 20, 130],
  [30, 160, 160, 30],
  [90, 90, 90, 90]
]

delay = 0.2

def map(value, from_low, from_high, to_low, to_high):
  from_range = from_high - from_low
  to_range = to_high - to_low
  scale_factor = float(from_range) / float(to_range)
  return to_low + (value / scale_factor)


def set_angle(channel, angle):
  pulse = int(map(angle, 0, 180, servoMin, servoMax))
  pwm.setPWM(channel, 0, pulse)

def dance_step(step):
  set_angle(0, step[0])
  set_angle(1, step[1])
  set_angle(2, step[2])
  set_angle(3, step[3])

def dance_pupet():       ❸
    for i in range(1, 10):
        for step in dance:
            dance_step(step)
            time.sleep(delay)

pwm.setPWMFreq(60)


while True:
  if GPIO.input(PIR_PIN) == True:      ❹
     mixer.music.play()
     dance_pupet()
     time.sleep(2)
```

❶ The first new bit of code is the code to set pin 23 to be a digital input.

❷ There is also some initialization code to start the mixer ready to play the sound clip.

❸ The new function dance_puppet is used to make the puppet repeat the dance steps 10 times.

❹ If the PIR sensor is activated (pin 23 is True), then the music track playing and dancing is started. The music.play function does its work in the background.

Using the Talking Puppet

You can record a new sound clip to play using the Audacity software that you may have used in "Arduino Experimentation" on page 290. Just replace the file *pepe_1.wav*. You could also really go to town on this project and record a number of different sound clips to be played at random, or customized to the time of day.

Summary

In this chapter, you learned how to use sound with an Arduino and Raspberry Pi.

In the final chapter of this book, we will explore the use of Arduino and especially Raspberry Pi in the Internet of Things.

The Internet of Things

In general, the primary way people interact with the Internet is by using a browser to view web pages. The *Internet of Things*, or *IoT*, is the concept that other "things" can also have a presence on the Internet. For example, smart appliances for home automation are connected to the Internet and can provide useful information to homeowners and utility companies. In addition, users of wearable devices (such as smart watches and fitness trackers) might consider themselves to be part of the IoT, as personal information about their location and heart rate finds its way onto cloud services on the Internet.

Controlling electronic devices takes on a whole new dimension if that control is over the Internet. In this chapter, you will learn how to take control of the actuators described in the book using a web interface.

In practical terms, there are two main ways of making your Raspberry Pi network and therefore Internet capable.

One is the direct approach to have the Raspberry Pi act as a web server. It can then serve up a web interface that you can connect to from any browser and interact with. So, for example, you press a button on the web page served from the Raspberry Pi and a GPIO output turns on. We will use this approach in "Project: A Raspberry Pi Web Switch" on page 301.

The second approach is for the Raspberry Pi to communicate with a cloud service that acts as a broker for messages to pass between devices and people on the Internet. For example, in "Project: Puppet Twitter Notifier" on page 305, the IFTTT (If This Then That) cloud service will monitor your Twitter account and make Pepe the Puppet dance every time someone tweets with the hashtag #dancepepe.

If you don't want to use Twitter, you can instead use email, Facebook updates, or pretty much any kind of trigger that the If This Then That service (*http://www.ifttt.com*) can use.

Raspberry Pi and Bottle

Bottle is a very lightweight and easy-to-use web server framework written entirely in Python. It's a great way of creating simple web server applications for a Raspberry Pi.

To install Bottle, enter the following commands:

```
$ sudo apt-get update
$ sudo apt-get install python-bottle
```

The next step is to create a Python program that uses Bottle to create a minimal web server that you can connect to from a browser anywhere on your network. Enter the following command to create a new Python file:

```
$ nano test_bottle.py
```

Then add the following text to the file:

```python
from bottle import route, run

@route('/')
def index():
    return '<h1>Hello World</h1>'

run(host='0.0.0.0', port=80)
```

The @route marker before the index function indicates that index is responsible for generating the HTML to be sent back to any browser that hits the root of the web server (the website itself rather than a particular page). In this case, it will just return the text "Hello World" formatted as a first-level heading. To check that the program works, run the following command:

```
$ sudo python test_bottle.py
Bottle server starting up (using WSGIRefServer())...
Listening on http://0.0.0.0:80/
Hit Ctrl-C to quit.
```

Then open a browser, either on the Raspberry Pi itself or on any computer in your network, and use the IP address of your Raspberry Pi as the URL (Figure 16-1). To find the IP address of your Raspberry Pi see "Finding the IP Address of a Raspberry Pi" on page 31.

Hello World

Figure 16-1 *"Hello World" in Bottle*

Project: A Raspberry Pi Web Switch

Because a Bottle web server is just a Python program, it can do more than just serve up HTML for a browser. It can also use the RPi.GPIO library to control the GPIO pins in response to hyperlinks or buttons being clicked on a web interface that it serves up to a browser (Figure 16-2).

You can use this device to turn AC appliances on and off if you use a PowerSwitch Tail, as described in "Project: Raspberry Pi Timer Switch" on page 263 in Chapter 13.

Web Switch

ON OFF

Figure 16-2 *A web interface for the Internet switch project*

Hardware

This project provides a web interface that will set GPIO pin 18 high or low from a browser. What you choose to connect to pin 18 is entirely up to you. You may decide to keep it simple and just use an LED. If so, build the hardware for "Experiment: Controlling an LED" on page 46. Or, you might have it control a PowerSwitch Tail and some AC appliance, as described in "Project: Raspberry Pi Timer Switch" on page 263.

Software

Bottle provides a neat way of keeping the HTML to be served to a browser separate from the Python program logic that controls the whole thing. The mechanism is called templating. This project uses just two files, which you can find in *projects/pr_web_switch/*.

The file *home.tpl* contains the HTML for the web interface:

```
<html>
<body>

<h1>Web Switch</h1>

<a href="/on">ON</a>

<a href="/off">OFF</a>

</body>
</html>
```

This HTML when viewed in a browser looks like Figure 16-2. The key lines are the two hyperlink a tags. The href attribute of the a tags specify the web address to go to if ON or OFF are clicked. If you are familiar with HTML and CSS styles, you may like to smarten up these hyperlinks and make them look a bit more like real buttons.

In the case of the ON hyperlink, a web request will be sent to the IP address of your Raspberry Pi with /on on the end of it. This is then handled in the Python code that will be discussed next:

```
from bottle import route, run, template, request
import RPi.GPIO as GPIO
import time

GPIO.setmode(GPIO.BCM)    ❶
CONTROL_PIN = 18
GPIO.setup(CONTROL_PIN, GPIO.OUT)

@route('/')        ❷
def index():
    return template('home.tpl')

@route('/on')      ❸
def index():
    GPIO.output(CONTROL_PIN, 1)
    return template('home.tpl')

@route('/off')     ❹
def index():
    GPIO.output(CONTROL_PIN, 0)
    return template('home.tpl')


try:
```

```
    run(host='0.0.0.0', port=80)   ❺
finally:
    print('Cleaning up GPIO')
    GPIO.cleanup()
```

❶   Set up pin 18 as the output control pin.

❷   If a browser visits the root of this web server, just return the contents of the *home.tpl* template.

❸   This is the handler if the URL path has on on the end of it (in which case, the control pin is set high before returning the home template again).

❹   The handler for the OFF hyperlink.

❺   Start the web server running on port 80 (the default port for web pages).

Using the Web Switch

To start the web server running, use this command:

```
$ sudo python web_switch.py
```

Then open a browser tab on the IP address for your Raspberry Pi, which should then display as Figure 16-2.

Try clicking the ON and OFF hyperlinks and whatever you have attached to GPIO 18 should turn on and off.

Arduino and Networks

With its built-in network interface and availability of low-cost USB WiFi modules, the Raspberry Pi is actually far more suited to IoT projects than an Arduino Uno.

You can buy add-on WiFi shields for the Arduino, but they are expensive. Other models of Arduino like the Arduino Yun include WiFi but are also somewhat expensive and not terribly easy to use.

If you want to use an Arduino-like WiFi device in IoT projects, then I recommend that you use something like the Photon board from Particle.io (Figure 16-3).

Figure 16-3 *The Photon*

The Photon was styled on a version of the Arduino called the Arduino Nano but has a built-in WiFi module. The device is a true cloud-based device in that you communicate with it and install software on it all over the Internet. This means that you can embed your Photon in a project and still program it without having to have physical access to it. It just needs to be powered up and connected to your WiFi network.

The programming language for the Photon is also based on Arduino C, but instead of the standard Arduino IDE, it is programmed from a web-based IDE. You can find out more about using the Photon at *particle.io* and also from my book *Make: Getting Started with the Photon*.

Another useful Arduino-ish device that also gets used in IoT projects is the ESP8266. These devices (Figure 16-4) are fantastically low cost and can be set up to program from the Arduino IDE as if they were an Arduino, or they can be connected to an Arduino to provide it with a low-cost WiFi connection.

Figure 16-4 *An ESP8266 module*

Using this device is gradually getting simpler, but (at the time of writing) it still requires a lot of configuration to properly set up. If you want to learn more about this device, then search the Internet for ESP8266 and try one of the many getting started tutorials that you will find there, such as *http://makezine.com/2015/04/01/esp8266-5-microcontroller-wi-fi-now-arduino-compatible/*.

Project: Puppet Twitter Notifier

Let's build on "Project: Pepe the Puppet Gets a Voice" on page 293 to make Pepe respond to tweets containing the hashtag #dancepepe, causing Pepe to do his little dance and play a sound. The project can use the exact same hardware as "Project: Pepe the Puppet Gets a Voice" on page 293, but with the PIR sensor removed. Only the software will change. Figure 16-5 shows the build from "Project: Pepe the Puppet Gets a Voice" on page 293 but with the PIR sensor removed.

Figure 16-5 *Pepe the Puppet wired up to respond to Twitter*

Putting Pepe on the Internet

Making Pepe respond to tweets is a two-stage process. The first stage is to allow Pepe to respond to web requests so that you can trigger his dancing from a web browser. The second step will use the IFTTT (If This Then That) web service to monitor Twitter for the #dan-cepepe hashtag and then cause a web request to be sent, to link up with the first stage.

You will use the dweet.io web service to get Pepe responding to web events. This service is free (subject to a reasonable limit on the number of messages you can send a month) and describes itself as Twitter for the IoT. It does not require a login and is very easy to use. It also has a Python library that will make integrating it with Pepe's software straightforward.

To install dweepy, the Python library for dweet.io, run the following commands:

```
$ git clone git://github.com/paddycarey/dweepy.git
$ cd dweepy
$ sudo python setup.py install
```

There is a slight problem using this library with Python 2 and SSL that is easily solved by using the following commands to change the version of HTTP requests that Python 2 uses:

```
$ sudo apt-get install python-pip
$ sudo pip install requests==2.5.3
```

You should now be able to run the program for this project (*puppet_web.py*), which you will find in *python/projects/puppet_web*, using this command:

```
$ sudo python puppet_web.py
```

You can test out the progress of the project with a web browser. Just navigate to the URL *https://dweet.io/dweet/for/pepe_the_puppet* (Figure 16-6).

Figure 16-6 *Controlling the puppet from a web browser*

As soon as you go to that URL, Pepe should start doing his stuff.

Much of the program is the same as "Project: Pepe the Puppet Gets a Voice" on page 293 so you may wish to refer back to that project for the main body of the code:

```
from Adafruit_PWM_Servo_Driver import PWM
from pygame import mixer
import time
import dweepy       ❶
```

```python
pwm = PWM(0x40)
mixer.init()
mixer.music.load("pepe_1.wav")

dweet_key = 'pepe_the_puppet'    ❷

servoMin = 150  # Min pulse length out of 4096
servoMax = 600  # Max pulse length out of 4096

dance = [
  #lh  lf  rf  rh
  [130, 20, 20, 130],
  [30, 160, 160, 30],
  [90, 90, 90, 90]
]

delay = 0.2

def map(value, from_low, from_high, to_low, to_high):
  from_range = from_high - from_low
  to_range = to_high - to_low
  scale_factor = float(from_range) / float(to_range)
  return to_low + (value / scale_factor)


def set_angle(channel, angle):
  pulse = int(map(angle, 0, 180, servoMin, servoMax))
  pwm.setPWM(channel, 0, pulse)

def dance_step(step):
  set_angle(0, step[0])
  set_angle(1, step[1])
  set_angle(2, step[2])
  set_angle(3, step[3])

def dance_pupet():
    for i in range(1, 10):
        for step in dance:
            dance_step(step)
            time.sleep(delay)

pwm.setPWMFreq(60)


while True:
    try:    ❸
        for dweet in dweepy.listen_for_dweets_from(dweet_key):    # ❹
            print("Dance Pepe! Dance!")
            mixer.music.play()
            dance_pupet()
    except Exception:
        pass
```

❶ Import the dweepy library.

❷ dweet uses this value as a key for dweets that are of interest you. If you leave they key like this, then other readers of this book who are working on this project will be able to trigger your puppet (and vice versa), which is kind of fun. If you want to make things more private, just choose a different value for this key.

❸ The code inside the main loop is all contained in a try/except error handler, because the web connection that the program listens on times out after a while, causing an exception. The try/except code masks this and allows the program to try again after such errors.

❹ dweet.io allows you to wait and listen for new dweets for your key and perform an action when they arrive.

IFTTT (If This Then That)

IFTTT is a web service that allows you to set up triggers that cause an action. For example, you could create an IFTTT recipe that sends you an email (the action) whenever someone mentions you on Twitter (the trigger).

In this project, IFTTT will be used to monitor Twitter for mentions of the hashtag #dance-pepe and then send the web request to get Pepe dancing and playing his sound clip.

Here are the steps involved in setting up IFTTT to do this job.

Step 1: Create a new recipe

Sign up to use the IFTTT service (it's free). Then click the Create Recipe button. You will then see the page shown in Figure 16-7.

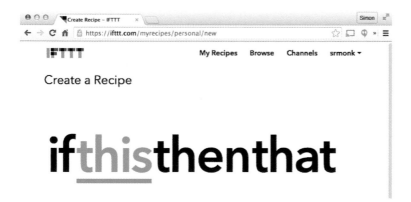

Figure 16-7 *Creating a new recipe in IFTTT*

Step 2: Define the trigger

Click the big "this" hyperlink and then find the Twitter channel from the list of channels. Within the Twitter channel, find the trigger "New tweet from search" and then enter "#dancepepe" in the "Search for" field (Figure 16-8).

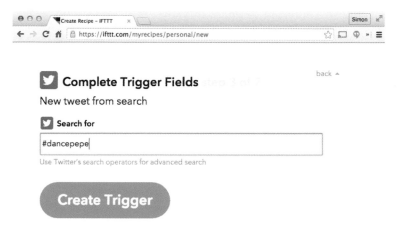

Figure 16-8 *Setting the Twitter trigger*

Step 3: Add the web request action

Once the trigger has been defined, the "if this then that" screen should reappear and now it's time to define the action by clicking the "that" hyperlink and then searching for the "Maker" action channel. Select the only action available ("Make a web request") and then complete the form shown in Figure 16-9.

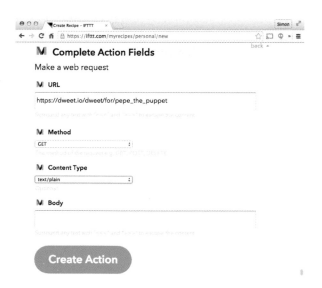

Figure 16-9 *Completing the action form*

Step 4: Finish the recipe

Click the Create Action button and then confirm the completion of the recipe by clicking Create Recipe (Figure 16-10).

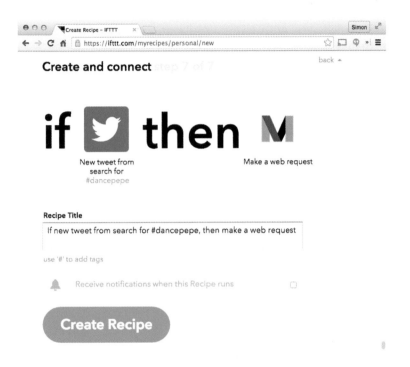

Figure 16-10 *Completing the recipe*

Once complete, the recipe will automatically go live. You may find that IFTTT asks you to log in to the Twitter Channel at some point in this process.

Using the Project

That's it, the project is ready to go. You can test it out by tweeting the hashtag #dancepepe. It may take a little while for IFTTT to find the tweet, so don't worry if nothing happens for a minute or two.

If, after a minute or two, nothing happens, you can check the IFTTT log for your recipe, which is the icon that looks like a list of bullet points on the page for the recipe. This will show you if the recipe was triggered (as well as any errors).

If you want to record your own message for Pepe to play, see "Arduino Experimentation" on page 290.

You should also explore the additional triggers available on IFTTT, as there is no end of interesting things that could set Pepe off!

Summary

In this chapter, you have learned how to run a web server on a Raspberry Pi and also make use of web services such as IFTTT and dweepy to make fun IoT projects.

I hope this book has served to provide you with some useful nuggets of information and some ideas for great projects that you can start making.

Parts

Finding parts for projects can be quite a challenge. In this appendix, you will find sources for the parts used in the book as well as extra useful information such as pinout diagrams for some components.

Suppliers

There are now many electronic component suppliers that cater to the maker and electronics hobbyist. Some of the most popular are listed in Table A-1.

You will find that most of the suppliers listed here sell the Arduino Uno R3 and Raspberry Pi 2 model B recommended in this book.

Adafruit's Arduino experimenter kit (product ID 170) and Sparkfun Arduino Inventor's kit (KIT-11227) are both a good way to get started with a basic selection of components and an Arduino. The MonkMakes Basic Components Pack has most of the resistors, capacitors, transistors, and LEDs used in this book.

Table A-1 *Suppliers*

Supplier	Website	Notes
Adafruit	*http://www.adafruit.com*	Good for modules
Digikey	*http://www.digikey.com/*	Wide range of components
MakerShed	*http://www.makershed.com/*	Good for modules, kits, and tools
MCM Electronics	*http://www.mcmelectronics.com/*	Wide range of components
Mouser	*http://www.mouser.com*	Wide range of components

Supplier	Website	Notes
SeeedStudio	http://www.seeedstudio.com/	Interesting low-cost modules
SparkFun	http://www.sparkfun.com	Good for modules
CPC	http://cpc.farnell.com/	UK-based, wide range of components
Farnell	http://www.farnell.com/	International, wide range of components
Maplins	http://www.maplin.co.uk/	UK-based, walk-in shops
Proto-pic	http://proto-pic.co.uk/	UK-based, stock SparkFun and Adafruit modules
Pimoroni	http://shop.pimoroni.com	Specialize in Raspberry Pi
MonkMakes	http://www.monkmakes.com	Component and Project Kits to support Simon Monk's books

Resistors and Capacitors

Resistors and capacitors are very inexpensive, but this often means that there is a minimum order quantity of 50 or 100 with some suppliers. It is often better to buy a mixed starter kit of components.

Description	Sources
10Ω 1/4 W resistor	Mouser: 291-10-RC
100Ω 1/4 W resistor	Mouser: 291-100-RC
150Ω 1/4 W resistor	Mouser: 291-150-RC
270Ω 1/4 W resistor	Mouser: 291-270-RC
470Ω 1/4 W resistor	Mouser: 291-470-RC
1kΩ 1/4 W resistor	Mouser: 291-1k-RC
4.7k 1/4 W resistor	Mouser: 291-4.7k-RC
10kΩ Linear Trimpot	Adafruit: 356 Sparkfun: COM-09806
Photoresistor	Adafruit: 161 Sparkfun: SEN-09088

100nF capacitor	Adafruit: 753
	Mouser: 810-FK16X7R2A224K
100uF 16V capacitor	Adafruit: 2193
	Sparkfun: COM-00096
	Mouser: 647-UST1C101MDD

Semiconductors

Components with part numbers like "2N3904" are easy to look up, but when it comes to finding common components like LEDs, it can be hard to find and it's usually best to look for an assorted kit of LEDs on eBay or Amazon.com or as part of a starter kit.

Description	Sources
2N3904 transistor	Adafruit: 756
	Sparkfun: COM-00521
	Mouser: 610-2N3904
MPSA14 Darlington transistor	Mouser: 833-MPSA14-AP
2N7000 MOSFET transistor	Mouser: 512-2N7000
TIP120 Darlington transistor	Adafruit: 976
	Mouser: 512-TIP120
FQP30N06L N-channel logic-level MOSFET	Mouser: 512-FQP30N06L
1N4001 diode	Adafruit: 755
	Sparkfun: COM-08589
	Mouser: 512-1N4001
Red LED	Adafruit: 297
	Sparkfun: COM-09590
Green LED	Adafruit: 298
	Sparkfun: COM-09650
Orange LED	Sparkfun: COM-09594
RGB LED diffused common cathode	Sparkfun: COM-11120

L293D H-bridge IC	Adafruit: 807
	Mouser: 511-L293D
L298N H-bridge IC	Mouser: 511-L298
ULN2803 8 x Darlington driver	Adafruit: 970
	Mouser: 511-ULN2803A
DS18B20	Adafruit: 374 (includes 4.7k resistor)
Encapsulated DS18B20 temperature probe	eBay, Adafruit: 381

Hardware

Breadboard and a selection of jumper wires are probably best obtained in one of the starter kits mentioned in "Suppliers" on page 313.

Description	Sources
400-point solderless breadboard	Adafruit: 64
Male-to-male jumper wires	Adafruit: 758
Female-to-female jumper wires	Adafruit: 266
Assorted hookup wire	Adafruit: 1311
Female-to-male jumper wires	Adafruit: 826
3V battery box (2xAA)	Adafruit: 770
6V battery box (4xAA)	Adafruit: 830
Female barrel jack to screw terminal adapter	Adafruit: 368
Two-way terminal block	Electrical/DIY store

Miscellaneous

A lot of the components in this section can be found on eBay and Amazon.

Description	Sources
Small 6V DC motor	Adafruit: 711
12V bipolar stepper motor	Adafruit: 324
5V unipolar stepper motor	Adafruit: 858
9g servomotor	eBay, Adafruit: 169
PowerSwitch Tail SSR	Adafruit: 268
5V at 2A power supply	Adafruit: 276
12V 1A power supply	Adafruit: 798
Power supply (12V at 5A)	Adafruit: 352
EasyDriver stepper motor controller	Sparkfun: ROB-12779
Adafruit 16-Channel 12-bit PWM/Servo Driver	Adafruit: 815
Small 8Ω speaker	Adafruit: 1891
WS2812 addressable LED strip	eBay, Adafruit: 1376
I2C OLED display 128x64 pixels	eBay
PIR sensor module	eBay, Adafruit: 189
Relay module	eBay

Pinouts

Figure A-1 shows the pin connections for some of the components used in the book.

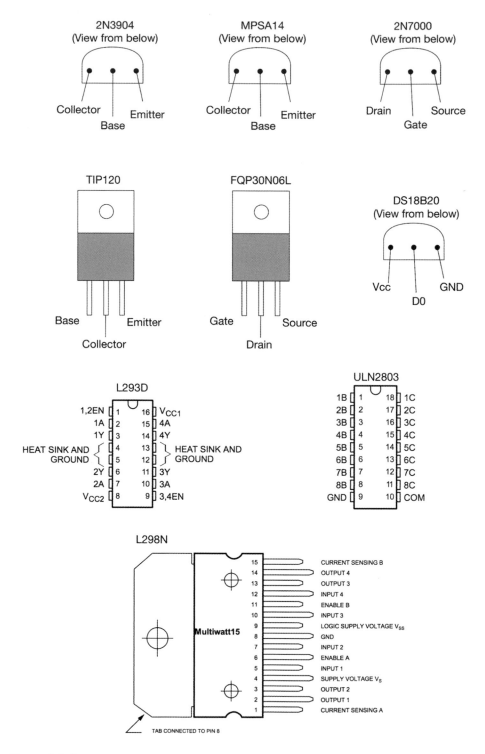

Figure A-1 *Component pinouts*

Raspberry Pi GPIO Pinout

B

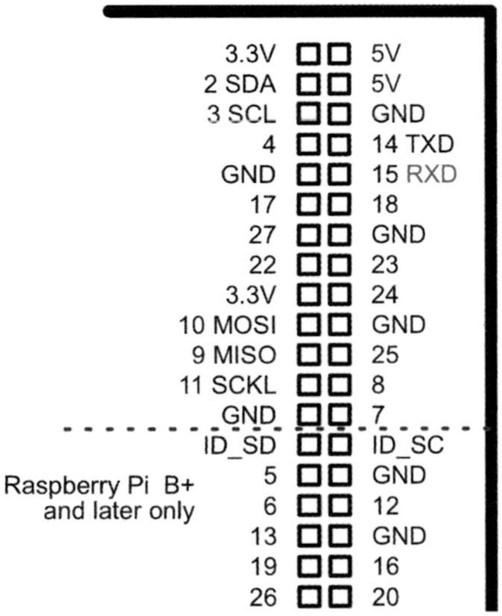

3.3V	☐☐	5V	
2 SDA	☐☐	5V	
3 SCL	☐☐	GND	
4	☐☐	14 TXD	
GND	☐☐	15 RXD	
17	☐☐	18	
27	☐☐	GND	
22	☐☐	23	
3.3V	☐☐	24	
10 MOSI	☐☐	GND	
9 MISO	☐☐	25	
11 SCKL	☐☐	8	
GND	☐☐	7	
ID_SD	☐☐	ID_SC	
5	☐☐	GND	
6	☐☐	12	
13	☐☐	GND	
19	☐☐	16	
26	☐☐	20	

Raspberry Pi B+
and later only

Figure B-1 *The Raspberry Pi pinout*

Notes

- I2C on pins 2 and 3
- SPI on pins 9, 10, and 11
- TTL Serial on pins 14 and 15
- Pins ID_SD and ID_SC are reserved for use with serial EEPROM for identifying add-on boards (HATs)

Index

Z

About the Author

Simon Monk writes books full time, mostly about electronics for Makers. Some of his better-known books include *Programming Arduino: Getting Started with Sketches*, *The Raspberry Pi Cookbook*, and *Hacking Electronics*. He also helps his wife, Linda, who runs MonkMakes.com, to make and sell kits and other products related to Simon's books. You can follow Simon on Twitter and find out more about his books at *simonmonk.org*.

About the Technical Reviewer

Duncan Amos has spent the majority of his 50+ year working life in TV broadcast engineering, but has also designed and built satellite subsystems, created technical handbooks and user guides, artificially inseminated farm livestock, repaired garden machinery, and designed and built furniture. Coming to microcontrollers late in life while being an inveterate Maker, he knows the value of clear explanation for complex techniques.

Colophon

The cover image is by Brian Jepson. The cover fonts are Benton Sans Bold, Benton Sans Regular, and Soho Pro Medium. The text font is Adobe Minion Pro; the heading font is Adobe Myriad Condensed; and the code font is Dalton Maag's Ubuntu Mono.